To Those Who
Have Confused You
to Be a Person

To Those Who Have Confused You to Be a Person

Words as Violence and Stories of Women's Resistance Online

ALIA DASTAGIR

Crown
An imprint of the Crown Publishing Group
A division of Penguin Random House LLC
crownpublishing.com

Library of Congress Cataloging-in-Publication Data
Names: Dastagir, Alia, author.
Title: To those who have confused you to be a person : words as violence and stories of women's resistance online / Alia Dastagir.
Description: First edition. | New York : Crown, [2025] | Includes bibliographical references.
Identifiers: LCCN 2024028897 (print) | LCCN 2024028898 (ebook) | ISBN 9780593727843 (hardcover) | ISBN 9780593727850 (ebook)
Subjects: LCSH: Women—Abuse of. | Women—Crimes against. | Women—Violence against. | Online hate speech. | Cyberbullying. | Harassment.
Classification: LCC HV6250.4.W65 D375 2025 (print) | LCC HV6250.4.W65 (ebook) | DDC 302.34/302854678—dc23/eng/20241127
LC record available at https://lccn.loc.gov/2024028897
LC ebook record available at https://lccn.loc.gov/2024028898

Hardcover ISBN 978-0-593-72784-3
Ebook ISBN 978-0-593-72785-0

Editor: Amy Li
Production editor: Liana Faughnan
Text designer: Aubrey Khan
Production manager: Dustin Amick
Copy editor: Janet Biehl
Proofreaders: Rachel Markowitz, Tess Rossi, Robin Slutzky, and Lorie Young
Publicist: Bree Martinez
Marketer: Chantelle Walker

Manufactured in Canada

9 8 7 6 5 4 3 2 1

First Edition

CONTENTS

To Those Who
Have Confused You
to Be a Person

Introduction

I'LL BEGIN WITH my daughter, with the first violation I clearly remember.

We lie in bed together, and I tell her it's time to sleep. She's unconvinced. Fading sun filters through the curtains, and she is brooding about all the play left undone.

Eventually, her body starts to succumb, one big toe rhythmically kneading the back of my thigh. A groggy arm eases across my belly, loose with skin that stretched to make space for her life. White noise shushes her room. I can sense the slowing of her breath, marveling at that peaceful descent into sleep. Dreaming worlds. Building worlds. I know when night deepens, she will leave her bed, walk soundlessly into my room, and whisper them to me.

I pull out my phone, open my email.

"Stop complaining cunt. You should consider suicide."

I look around the room at the markers of my daughter's nascent life. I don't understand how these words got into this place. A drawing of a portly turtle, a *Le Petit Prince* music box, a polka-dotted sock strewn without its mate. I place my palm on the small of her back, devastated by the gesture's futility.

I am afraid. Of what? Words? I am a journalist. The reader is upset by a story I wrote. I need to lock the door. It's a normal thing to do.

• • •

More than a year later, my daughter has grown inches, and her new baby sister sleeps in my arms. I rock the baby in a velvet chair, in a room that smells of beginning. The only language she knows is the language of bodies. My heart in her ears as she swam inside me, a cheek at home across the slope of her father's shoulder, her tiny fingers tracing the swollen blue rivers along her grandmother's hands. In the rocking chair, our bodies speak this first language.

I open my Facebook Messenger. Another reader, angered by a story I wrote:

"I am sorry to those who have confused you to be a person, because you are not a person."

I gaze at this fresh baby, thumb her cheek, touch her feather-fine hair. I think to myself: What indignities will the world have you suffer? When you learn the language of a world maintained by violence, what words will you hear when people try to silence you?

My baby stirs. I feed her from my body.

• • •

Two years pass, and then the mob comes. Thousands are enraged over a story I wrote. Naïvely, I don't expect it. QAnon unleashes them. Donald Trump, Jr., encourages them.

For days, they beat me with language. Conservative bloggers, far-right firebrands, conspiracy theorists. They tear at me from behind their avatars—voluptuous anime, Pepe the Frog, men beside grinning girlfriends, a mother at Disney World.

There are death threats—to put me in the wood chipper, to shoot me. They say they would not feel sorry for my children if they were raped.

The baby is a toddler now. She sees me on the stairs after hours holed up in my room, curtains drawn, in the stale semidarkness. "You miss me," she says.

Editors encourage me to call the police, and now the officers are on my doorstep—the second time in six months. I sob in my closet, my guts a bundle in my throat. My husband stands outside the closet door.

"It's not worth it anymore!" I scream.

He doesn't know what to say.

In the years before the mob came, I would occasionally read the abuse to my mother, who laughed and sometimes gasped, but this time they found her and threatened her, too, and the fear made her quiet.

When Trump Jr.'s conservative army advanced upon me, when I became that day's target around which they could collectively define the contours of their rage, I didn't believe anyone could see that my mind was on fire, that I worried my anxiety would kill me, that my body was wrecked by terror.

I was flesh and bones and muscles intact, but inside my nerves were kinetic. I was nauseous for days, reminded of both my pregnancies, except this sickness was caused not by hormones signaling cells in my womb to multiply into life but by stress hormones whose frequent resurgence was sending my own cells into early death. Paranoia curled around each thought. Could they find my search history? Read my texts? My snark on Microsoft Teams? What evidence could they find of me being a bad journalist, a bad woman? After manic interrogation, there was flatness. Their hate leveled me.

After two days, the mob changed direction, shifting its sights toward fresher targets. I remained still and unsettled. I would pass out in the middle of the workday, folded in a posture of defeat, in between interviews I was too exhausted to prepare for.

In the month after, I put myself back together in pieces: hygiene, presence, productivity, hope. I wondered how fluent my body was becoming in these adaptations. Six weeks after the mob's assault, I went to Costa Rica. At night, monkeys darted along roofs making sounds like showering hail. In the morning, everything was buttery light. I don't usually do yoga, but that day I tried. My forehead kissed the green mat, pressed against the spongy grooves as ceiling fan gusts caressed my spine. The yoga teacher, whose body was a tapestry of ink and story, said, "Here we are safe," and I felt grief beneath my eyelids. I thought about how often it feels that I am not.

• • •

For nearly half a decade when I was a reporter for *USA Today,* online violence changed the texture of my life. At first I did not call what was happening violence. I didn't know what to call it.

We have heard this problem called many words: online abuse, online harassment, online hate. We have heard it called an old problem and a new problem. Sometimes we don't name it. Maybe because it's so bad, the names we have don't feel like enough, or maybe because we think what happened to us is not so bad, compared to what has happened to other women.

I have listened to women talk about these experiences using words that feel right to them: I was harassed, I was abused, I was terrorized. Without reservation, I now tell anyone who will listen: I experienced violence.

Violence researcher Sherry Hamby wrote in her paper "On defining violence, and why it matters" that "violence research has produced numerous and sometimes conflicting definitions," but that "a comprehensive definition . . . includes 4 essential elements: behavior that is (a) intentional, (b) unwanted, (c) nonessential, and (d) harmful."

When I am with my child and someone hurls slurs into my phone, into her room, into my body, when we should be soft and still together, when someone's vicious syllables turn the roof of my mouth heavy and my stomach into stone, when they make me wonder about my words and my worth, when they do this again and again at cost to my health and my dignity, when they spread lies about who I am and what I do that compromise my safety, when they threaten my life and make me fear for the lives of people I love, when I am powerless to stop it, this is violent.

In 2019 I decided to write a book about women and online abuse. I did not have all my language yet, but I was bewildered by the facts of my experience. I was especially confused by the message—tacit and explicit—to "ignore it," which I think sometimes suggests *do not react*, and other times suggests *do not feel.* When women report these violations and social media companies do nothing, the message we get is that if we want to remain on these platforms, we have no choice but to ignore it. When we call the police to report threats and they do not act, the message women get is that we have no recourse, so we should get a dog, or a gun, or otherwise ignore it. When women share publicly about their abuse and people tell us, "This is just the Internet," the message we get is that everyone else is ignoring it, and if we want to be online, we need to ignore it, too. When well-meaning people in our lives tell us not to let it get to us, to keep going, to keep posting, to keep speaking, to keep publicly thinking, the suggestion is that for our own good, we must ignore it.

I could not ignore it. I suspected, no matter how inured a woman became, "ignoring" was not the right word for how women were surviving violence.

Women of the blogging era and women who were among the earliest adopters of social media have been trying to name this violence for so long, but tech leaders who had the power to do something about it didn't listen well or they heard very clearly but did not care. Black women, who are disproportionately targeted for online abuse, were vocal in trying to call attention to online harassment long before it entered mainstream consciousness, but their concerns were minimized and their experiences and work on the issue continue to be erased. Black digital feminists and technology experts like Sydette Harry and Shireen Mitchell were documenting harmful and abusive practices online, including white supremacists masquerading as Black women, in the early 2010s. White supremacists were merciless after the killing of Trayvon Martin in 2012, sending Black users photos re-creating Martin's dead body.

Black feminists were also among the first to see the alt-right at work. In the summer of 2014, Harry and fellow Black feminists Shafiqah Hudson and Ra'il I'Nasah Kiam created the hashtag #YourSlipIsShowing to expose a campaign against trolls pretending to be Black women, which showed how an antifeminist propaganda operation used fake Twitter accounts to manipulate discourse and sow discord. While Harry, Hudson, and I'Nasah Kiam led the effort, the hashtag was used by a community of Black Twitter users to collectively generate data, ask for help, and expose 4chan members active in the campaign. Many of the people who called out the imposters became targets themselves.

Months later we got Gamergate, an online backlash against women game developers and feminists critiquing video game culture, spawned by the same bad actors who had coordinated in the same corners of the Internet to impersonate Black women online. Gamergate is considered a watershed cultural event, an implosion of masculine right-wing aggression toward feminism that demonstrated the breadth and severity of organized online harassment campaigns. Mainstream media coverage of Gamergate centered white women victims, largely framing the problem as exclusively about misogyny and ignoring the campaign's links to a broader culture of abuse against marginalized people online. Gamergate became a huge media story. Then it became a harbinger.

Where are we now? Sydette Harry, who has been archiving violence online since the earliest days of blogging and social media, told me we have so many solutions to mitigate harm but that those best positioned to enact meaningful change remain largely intractable.

"Is it unsolvable? Absolutely not," she said. "We've done more difficult things in less time. . . . We lack will."

It is difficult work to continue insisting that your words and experiences matter. Even now women who speak about violence online often find their words thwarted by ignorance, malice, and greed. As if the abuses we named were not enough, our naming is turned against us: Fragile. Sensitive. Whiners. As feminist scholar Sara Ahmed writes in her book *Complaint!*, "To be heard as complaining is not to be heard. To hear someone as complaining is an effective way of dismissing someone. You do not have to listen to the content of what she is saying if she is *just* complaining or *always* complaining."

When journalist Taylor Lorenz said during an interview that her own online abuse got so bad that she wanted to kill herself, when she cried and pleaded, her abusers turned her agony into a meme.

• • •

This is a book about how women are surviving violence online. I began by wanting to be more attentive to the less visible ways in which pain and despair are managed and communicated. The subjects in this book are diverse, because while our experiences share certain features, our responses are rooted in our identities, our histories, in those many unknowable details that make a life. Toward the end of my reporting, I was moved by the determination and cleverness of my subjects. I was also struck by the impossibility of our situations. Sometimes the way we attempt to make ourselves feel emotionally better in the moment may not be what ultimately animates us to change the conditions of our lives.

This book documents what it means to be a woman online. It is an attempt to slow down a conversation, to eschew equivocations about which violations matter. It is written for women, but it should interest anyone who cares about how the Internet is shaping life.

You may draw strength from these women's stories. You will see the benefits and limitations of conscious and unconscious strategies women use to defend themselves against violence, including emotion regulation, humor, anger, numbing, and confrontation, among others. My simple hope for this book is that it will deepen women's understanding of their experiences online and show them why abuse in digital spaces is impossible to ignore. My loftier goal is that after reading, you will feel moved to demand better, to demand more, that you will feel compelled to insist on reforms that will improve the search engines, the platforms, the technology, the law, the culture.

As radical feminist philosopher Marilyn Frye wrote in her essay "Oppression," "One of the most characteristic and ubiquitous features of the world as experienced by oppressed people is the double bind." In the absence of structural reform, all our options are insufficient, and all exact some penalty.

I do not believe pervasive violence online is immutable, but in the face of it, I believe our options for coping reflect the absurdity of our situations. Humans built the Internet, we imbue it with meaning, we decide how it works and what it should do. My hope is that by the time you are finished with this book, you will have disabused yourself of the punishing idea that your only option for surviving online is to accept violence as inevitable. We can imagine something different. I believe we must.

• • •

Online abuse is a tool used by people of every gender and across the ideological spectrum. There are many reasons for it: money, politics, outrage, norms of behavior online. This book is concerned with online abuses that reproduce hierarchies of power and perpetuate oppression. It leans toward gender-based violence perpetrated by men, but because oppressive ideologies are internalized by women, it explores women as perpetrators, too. In some places, it touches on intra-community harassment.

The Pew Research Center found that the number of Americans reporting severe harassment increased from 15 percent in 2014 to 25 percent in 2020. According to Pew, 61 percent of women say online harassment is a major problem, but online abuse doesn't affect just women. Pew found that overall men are somewhat more likely than women to say they have experienced any form of harassment online, though women are more likely to report being sexually harassed or stalked. About half or more Black or Hispanic targets say they were harassed online due to their race or ethnicity. Hate speech and harassment online are a significant problem for people who identify as LGBTQ.

There are important shades of difference between attacks waged against men, especially white, heterosexual men, and attacks waged against women and other marginalized people. For women, this abuse mirrors the discrimination we experience offline, it is often sexualized and constructed to humiliate us, it aims to push us out of spaces where men view us as competition, and it connects to larger structural violence.

Marilyn Frye wrote that it is the human condition to experience suffering and limitation, to encounter barriers, to be damaged, but that oppression is specific, an experience of living that is "shaped by forces and barriers which are not accidental or occasional and hence avoidable, but are systemically related to each other in such a way as to catch one between and among them and restrict or penalize motion in any direction. It is the experience of being caged in" by barriers that are "erected and maintained by men, for the benefit of men," even when sometimes those barriers may create difficulty and inconvenience for men, too. To determine whether a barrier is contributing to a person's oppression, she encourages one to ask, "Who constructs and maintains it? Whose interests are served by its existence? Is it part of a structure which tends to confine, reduce and immobilize some group? Is the individual a member of the confined group?"

When women are attacked online, the goal, conscious or not, is often to depress our participation on the Internet, and more broadly in public and civic life. These attacks come from many places. It is not just trolls, which digital scholar Whitney Phillips writes claim to be "in the service of nothing but their own amusement." Calling everyone who harasses women a troll is a misnomer and a minimization and a way to cognitively contain the problem. Not everyone targeting us is doing it for laughs. Some have political motivations, others have moral concerns, many have economic incentives, and not everyone is anonymous. Some of the people harassing us happily attach their names to their messages.

Rather than focus on the identities of individual perpetrators, it may be more useful to think of the problem as animated by ideologies. Anyone can internalize sexist, racist, white supremacist beliefs, and the people most likely to be targeted online are those who challenge them. As researcher Alice Marwick writes, while online harassment is a tactic used by people with diverse ideological and political leanings, it is "linked to structural systems of misogyny, racism, homophobia, and transphobia, which determine the primary standards and norms by which people speaking in public are judged."

I considered writing a more expansive book exploring violence toward all marginalized people online, but I ultimately concluded that the experiences of women's online abuse were distinct enough to warrant specific attention. I also felt that because I am a woman, that is the experience I could speak to most intimately and with the most authority. The experiences of other marginalized groups online deserve attention, too.

This book is inclusive of all those who identify as women. As Minnesota representative Leigh Finke, the first openly transgender person to serve in the state's legislature, wrote about online abuse in her blog: "The hate . . . I endure reminds me just how real my womanhood is."

I interviewed more than three dozen women whose experiences of harm are compounded by their racial, cultural, gender and sexual identities, religious identities, age, class, and disability status. I also interviewed dozens of experts and activists working on the problem of online abuse, and dozens more who were not but whose expertise in neuroscience, linguistics, philosophy, psychology, sociology, and healing enhances our understanding of the violence we face. I found a tremendous amount of psychological damage. I also found rich, creative, culturally specific ways of resisting. I had access to as much of these women's interior lives as they would allow, which means, of course, there were parts they kept for themselves.

I focused on adult women, as the abuse of young people online has unique dynamics that are beyond the scope of this work. This book does not delve deeply into how abuse and harassment differ by social media platform. These differences matter, as it is the technical affordances of a platform that enable or discourage abuse. But technology changes rapidly. I did not want to tie this book to the presence or absence of elements that are always in flux. And apart from the story of an Indigenous woman living in Canada, this book exclusively features stories of women living in the United States. Gender-based violence is a global problem, but its severity and texture, as well as its associated risks, vary from country to country, culture to culture.

There is no doubt that the framing of these stories is influenced by my own background and beliefs, which shaped my questions as well as what I found most relevant. These narratives were curated by a feminist, a straight woman, the daughter of an Afghan immigrant and a white mother, by a journalist who could afford to leave her full-time job to write these pages. I have tried to listen well, to approach the experiences of my subjects with careful curiosity and sensitivity, though there are many aspects of their specific burdens I will never fully comprehend.

I will acknowledge I have made many mistakes on my way toward this book. I grew up in a community whose politics I had to shed. I came into my feminism slowly, publicly. You can see it in my journalism. Sometimes the work was good, other times it was too tentative, too glib. In hindsight, there are some stories I wrote that likely perpetrated harm, or at the very least made someone else's harm invisible. I platformed people who didn't deserve it.

Like this book's narrator, the women interviewed are not perfect. You will not agree with everything they have written or said or done. You will not agree with all of their behavior online. You do not have to. You do not have to like or admire them to condemn their abuse.

I am writing this book because documenting harm is a moral imperative, because sometimes adding testimony to the record can change policy and law and public perception. This book is concerned with women's reflections on how a wide range of behaviors online impacted their lives—what it made them feel, what they lost, how it reshaped them. In this book I forgo a precise definition of online abuse, because I want women to be able to describe experiences that might not always fit within a popular definition. I use the terms *online violence, online abuse,* and *online harassment* interchangeably, sometimes to reflect how a woman spoke about her own experience, and because this is true for how we speak about the problem in the culture.

When I told feminist philosopher Kate Manne that my book rejected a rigid definition of online abuse, she told me my choice reminded her of the philosophical concept of family resemblance, which suggests that things you might think are connected by a single feature are actually connected by overlapping similarities, much as family members might share similar noses and chins but have different hair and eye color. You can see resemblances, bits of convergence, but it's not a neat, perfect tie-it-up-in-a-bow phenomenon. This book centers harm, and while some of my examples will have a set of features that are easily describable—explicit misogynist abuse, for example—others will document violations that are harder to form perimeters around. Centering harm allows for differences and similarities. It allows us to define for ourselves what harm has been committed against us.

• • •

To consider the present the book explores, we must briefly visit the past.

In 1996 cyberlibertarian John Perry Barlow warned against government regulation online in "A Declaration of the Independence of Cyberspace": "Ours is a world that is both everywhere and nowhere, but it is not where bodies live. We are creating a world that all may enter without privilege or prejudice accorded by race, economic power, military force, or station of birth. We are creating a world where anyone, anywhere may express his or her beliefs, no matter how singular, without fear of being coerced into silence or conformity."

This highly optimistic and utopian vision was also propounded by academics, journalists, and corporations. In the 1990s an MCI advertisement declared that on the Internet, "there is no race, there are no genders." Online, they told us, we could be anyone we wanted to be.

It's laughable now. The Internet affords degrees of anonymity, but attitudes about race, gender, and class are everywhere online. A raceless, genderless Internet was not even what many people wanted. As sociologist Jessie Daniels writes in her book *Cyber Racism,* people do not go online "in search of some disembodied libertarian utopia." Many people are online to affirm who they are, to be seen and heard, and to find others like them. Many women are online to build movements, to narrate their own stories, to self-define. We can't separate these activities from our bodies. Our bodies are how we think, how we feel our way toward something different. It's by listening to our bodies that we continue to survive.

The Internet as we experience it today was largely envisioned by men, most of whom were white, and it has continued to evolve with their perspectives. Discrimination is baked into the infrastructure of the Internet and its values. These are not neutral systems. Companies like Google, Facebook, and X (formerly known as Twitter)* may romanticize themselves as sites of knowledge and connection, but their primary objective is profit, which is why they have developed algorithms to keep us clicking and engaged, without enough care around what content we engage with, whether it is true, or how it causes harm.

Content moderation has been put forth as one solution to mitigate the harms that technology created. Content moderation uses artificial intelligence as well as human labor to reduce the amount of toxicity and abuse online. But this means that while some harmful content may not always reach the end user, another human being often has to engage with it, and that worker is typically paid wages that cannot possibly justify exposure to such content, if there is any justification at all.

This is not to say that nothing is improving. Caroline Sinders, a machine-learning-design researcher and online harassment expert, said there is greater literacy of digital harm today than there was a decade ago. Scholars like Danielle Citron are pushing for a cyber civil rights legal agenda, while lawyers like Carrie Goldberg have made gains in achieving legal accountability for men who engage in harmful conduct online, including revenge porn. Many employees who work in Big Tech are pushing for change and accountability. In

---

*Throughout the book I refer to the platform as Twitter because most of these stories occurred when the platform was still under this name.

2018 staff at Google offices around the world walked out over the company's mistreatment of women employees. In 2020 Facebook staffers staged a virtual walkout to protest the company's decision not to take action against posts by then president Donald Trump that appeared to threaten violence against protesters in Minneapolis after the police killing of George Floyd.

But many obstacles remain. Across social media platforms, trust-and-safety teams, whose job is to keep users safe from harm, are being deprioritized through staff reductions. In late 2022 tech billionaire Elon Musk bought Twitter, rebranded it X, reinstated many accounts that had been banned for hateful, violent rhetoric, and dissolved its Trust and Safety Council. Since then, Musk has used his platform to promote conspiracy theories and became a major backer of Trump's 2024 presidential campaign. In the United States and around the world, there has been a rise in authoritarianism, which contributes to digital harm broadly and online gender-based violence specifically. The COVID-19 pandemic put us online in new and enduring ways. The advancement of AI is threatening to pollute and overwhelm the Internet with more racist and sexist content. Disinformation is damaging women's reputations, credibility, and the integrity of our political process. Internet subcultures are manipulating the media into further abusing women while influencers build empires through the monetization of misogyny. After the 2024 election, there are legitimate fears that Donald Trump's return to power presents a grave danger to free speech, as many believe he plans to actively suppress dissent.

Harvard legal scholar Alejandra Caraballo told me the individual harms of harassment online are still widely misunderstood, and the social harms are frighteningly underestimated.

"It is going to be the way that political decisions are effectuated at all different levels of government," she said of online abuse. "It's corrosive to democracy."

• • •

In 2017, I interviewed Gloria Steinem in her home, and I share the story of that interview to show you where this book began.

Most of my reporting at the time of the interview focused on everyday women. It was March 2017, mere months after the first election of Donald Trump. Hundreds of thousands of people had poured into Washington, D.C., for the Women's March. Women were terrified of the violence we knew was ahead, and we were looking back toward history for knowledge, for help, for hope. Steinem, who had raised her voice against gender-based violence for more than half a century, told the swollen D.C. crowd, "We are here and around the world for a deep democracy that says we will not be quiet."

It was New York twilight when I arrived to interview Steinem at her home on the Upper East Side, though I'd been pacing nearby for nearly two hours, muttering self-talk. A less anxious videographer eventually joined me, and we entered the walk-up together. Inside, time shifted. Magazines and mementos blanketed walls and cluttered armoires: the famous black-and-white photo of Steinem and activist Dorothy Pitman Hughes, fists raised; Steinem and her friend Wilma Mankiller, who became the first female principal chief of the Cherokee Nation; a signed photo of Steinem and former president Barack Obama. Piles of paper coated her desk, whispering of a life's work that later generations endeavor to complete.

Shortly after we arrived, Steinem, then eighty-two, carefully descended the staircase that led from the second floor to the level where we had set up. I sat, calmed down, and began the interview. I had prepared what I would ask, when I would smile, how I would hold my pen. She seemed to be performing too, keeping the focus where she wanted, delivering lines she'd spent a lifetime repeating. When it was over, I was flooded with relief. I began to gather my things.

"You prepared for that," she remarked.

I suppose I felt then that I could speak without pretense. Words tumbled out about my fears, my frustrations, a burgeoning feminism. I told her about the story I wrote for Women's History Month on the parallels between the 2017 Women's March and the historic 1977 National Women's Conference, which she had attended. We were still demanding the same things: affordable childcare, equal pay, access to safe, legal, affordable abortion, freedom from all forms of violence. I told her about the hate in my inbox.

"You're reporting it, right?" she asked.

I was flustered. Back then, it hadn't occurred to me to do anything about my abuse. I thought I had no choice but to accept it.

She told me I had to report it. Otherwise, she said, "they get away with it."

Her advice was tactical, and one could argue that it overstated the effectiveness of online reporting mechanisms, but her specific instruction was not what I left with. It was the larger idea.

Don't ignore it.

I took that notion with me that day, evidence that when women commune, we birth novel ideas, new movements, different refusals, fresh hopes.

There is no right way to process hatred, so the advice to ignore it can sometimes seem reasonable. Fuck the haters. Toughen up. Thicken your skin. When a stranger fires off an email filled with horrible things aimed at your life. When someone on Twitter says you would be better off dead. When they spread lies that harm your reputation, your job, your bank account. When you go onto a stage, when you walk to your car, when you shut off the lights for the night, when you wonder if this person is going to murder you. Ignore that, too.

But who does this advice benefit?

In 2020 I shared some of my first writing on this topic during a workshop on personal disclosures led by author Roxane Gay, who told me she resents the expectation that anyone should be asked to ignore violent language, to remain stoic and steadfast in the face of hate. But she still admitted to me, "A lot of times we don't know what to do with it."

As Gamergate target Zoë Quinn wrote in their memoir *Crash Override*, "I can help someone secure their accounts in my sleep. I can sniff out and help remove dox like some kind of weird internet bloodhound. . . . The hardest part is helping someone cope."

Coping is how we work through negative emotions. It's our capacity to soothe, comfort, regulate, and calm ourselves. Sometimes it's how we combat lies about ourselves. Susan McGregor, a research scholar at Columbia University's Data Science Institute whose work centers on security and privacy issues affecting journalists, told me that she'd had an aha moment after she began learning about psychological first aid and other nonclinical mental health interventions, like peer support. She saw that severity and duration of online abuse were not nearly as important to long-term outcomes as social support and connections. So instead of asking targets about the harassment itself, she started to ask, "What helped?"

Each chapter of this book opens with a woman's story of violence online, explores the larger psychological, social, cultural, and political questions that arise from her story, and examines a specific way that woman coped with her abuse. Other women's stories are also woven in, along with my own reflections.

When I consider what has helped me survive the challenges of my own life, it wasn't a single strategy but a better understanding of myself. There is evidence that in some contexts, understanding what the pain is, where it's from, and that it's not your fault can reduce the experience of the pain itself.

When I interviewed Dr. Judith Herman, who has spent her career studying trauma victims, she told me the concept of this book reminded her of her friend Kathie Sarachild, whose consciousness-raising movement in the 1960s and '70s brought women together to talk about their feelings and experiences of being women as a method of political organizing. Sarachild believed that knowing more about one another would make women better equipped to fight on behalf of one another.

When we ask women to ignore violence online, what we are saying is that it is fine to live in a world where it's considered normal, acceptable, and legitimate that people abuse one another all the time, particularly people who are in the nondominant class. But we created these technologies, we created language, and we created the boundaries around both, which means we can draw new ones. When women ignore violence, when we don't talk about it or acknowledge its harm, we have no place to put it except inside ourselves.

CHAPTER 1

Just Ignore It

ALEXANDRIA ONUOHA IS on a bright spare stage dressed in white, bare feet, her black hair slicked tightly back. She is kneeling, but when the music begins, she quickly rises, arms eager and legs unbound. Her joints share a smooth vocabulary. She is soft wrists and loose limbs, blooming bones and fluid hips. She dances from the inside.

I ask her, What does it feel like to dance?

"Like everything makes sense," she says.

She lingers there, speaks of history, family, Blackness, womanhood. I count one, two, three, four times she tells me:

"I feel free."

Alex is a dancer, but there were times when people did not think she moved like one. She was a Black woman studying dance at a small, predominantly white liberal arts college in the Northeast. She moved her body according to the instruction she was given, but she often felt stiff and mechanical. "Robotic," she said. The dance genres she grew up in, the languages her body spoke easily, were hip-hop, liturgical, West African, dancehall. She used movement to fuse culture and art, sexuality and spirituality, past and present.

Alex struggled in her dance program, and she recounted to me the aftermath of going public with her experience. She had been unsettled by a white male guest dancer's comments on her body throughout rehearsals: the way it was failing, the way it did not fit.

She doesn't remember precise words, but she remembers his tone registering as sarcasm.

Alex spoke to a professor about the guest dancer's comments, but she did not feel she was taken seriously. When she got her grade, it was less than she believed she deserved, and she brought it up again to her professor, who she says dismissed her, telling her, in substance, "Sorry, this is just dance." Alex didn't think it was just dance.

Other professors made comments that suggested her body didn't belong, and seemed baffled by the way it moved. She didn't know what to call these critiques. Professors and guest dancers said they didn't understand what her body was doing or what her art meant. She produced a choreographic piece combining dance from her Jamaican and Nigerian roots. When it was time to perform it, a guest artist said, "I don't really understand what your piece is about, like, I am kind of confused, like, what's the point of having Bob Marley speak?" She again told a professor she felt the comment was not right. The professor said the comment was fine. She told Alex to grow thicker skin.

Alex tried, but near the end of her program, she was exhausted. She was exhausted by the side-eye, the erasure of Black art, and what she saw as favoritism of white bodies. She decided she needed to speak. She wrote an opinion piece for her school newspaper on what she experienced in her program. She called it "Dancing Around White Supremacy."

When the article ran online, friends saw it and texted to say they were proud. But at night, when she got back from an event and logged on to Instagram, she saw the other messages. She read them alone in her room.

She did not cry. She was still. She thought: "I can't believe I go to school with people who think like this."

She didn't recognize names, and not everyone used avatars, but she assumed the DMs were from other students. Who else, she thought, would read her school newspaper? In a school with a student body of less than two thousand, she imagined the messages were sent by people she ate with in the dining hall, sat with in class, passed on the way to her dorm. Online, they called her a "black bitch." They called her a "n—."

The day after the op-ed was published, Alex had dance class. She walked into class with dread in her step. She felt sweat coat her back. She didn't want anyone to know what was happening inside her body, so she made it unreadable. She disciplined her face.

When class began, Alex said, the professor didn't talk about Black art. She didn't talk about Alex, how she felt, what she and the other students of color needed. Instead, she suggested that students should be careful, especially with what they say about guest artists. Someone could get sued, Alex remembers the professor saying.

After class and the professor's not-so-subtle chiding, some of the white women dancers from her class came up to her in the cafeteria and said, "Oh Alex, we appreciate you being so courageous." But Alex said none of them spoke in front of the professor.

Members of the school administration met with her after the op-ed was published. They said her choice could follow her, and they wanted to make sure she understood all the potential ramifications for her future academic career.

They did not say the words *perhaps silence will keep you safe*. But that was what she believed they meant.

White Supremacy

The problem of women's online abuse is almost always framed as a problem of misogyny, but Alex's story, and the stories of countless other women, show that violence online is also linked to a struggle over the structural power of white supremacy, to the other systems with which it intertwines. Alex could not ignore what people were saying to her online, because language can be used to maintain power or to resist it. It can be used to keep certain people in their place or to fight a system that ranks human life. Language carries long histories and deep hopes. Our brains and our bodies experience language, reacting to generosity or to malice. Language influences how we see ourselves, how other people see us, how they treat us. Language shapes public life. So do silences.

When Alex was abused online, she was punished through multiple attack vectors: her gender, her race, and the norms of behavior for Black people in predominantly white spaces.

Black women online experience what feminist scholar Moya Bailey termed *misogynoir*, an anti-Black racist misogyny that produces a "particular venom." Research by the UK charity Glitch analyzed social media posts from July 2022 to January 2023 and found over nine thousand more highly toxic posts about Black women than white women, rife with stereotypes "such as 'the angry Black woman' ('Sapphire'), fetishisation ('Jezebel'), and fatphobia ('Mammy')." Black women journalists and politicians are 84 percent more likely to be mentioned in abusive or problematic tweets than are white women journalists and politicians, according to a 2018 report from Amnesty International. A 2023 study led by Michael Halpin of Dalhousie University in Halifax, Nova Scotia, found that women of color are "doubly denigrated" by the incel community "through a combination of racism and sexism."

Before the 2014 harassment campaign dubbed "Gamergate" became a cultural inflection point for the issue of women and online abuse, Black women were already navigating rampant misogynoir online. Gamergate was an explosion of masculine aggression toward women game developers, feminists critiquing video game culture, and anyone who dared defend them. Trolls organized on forums like 4chan and Reddit and the text-based chat system IRC to spread lies and disinformation about women they did not like, and they used those stories to justify attacks. Alice Marwick, principal researcher at the Center for Information, Technology, and Public Life, told me that during Gamergate, trolls "were looking for proof of concept," and they discovered their strategy worked: "It shuts people down, it gets them to stay off the Internet."

Most mainstream coverage of Gamergate focused on misogyny as an animating force, neglecting a deeper interrogation of the way racism also shapes the experiences of women online. Black women are gamers, too, and savvy Black digital feminists had already documented the harmful behavior of 4chan users who coordinated to impersonate and harass Black women online. Just months earlier, Shafiqah Hudson, Ra'il I'Nasah Kiam, and Sydette Harry had created the hashtag #YourSlipIsShowing, a nod to Hudson's Southern roots, letting the trolls know "we see you." Scholar Jessie Daniels, an expert on Internet manifestations of racism, told me the cultural conversation around Gamergate flattened the race element. White supremacy online, she said, does not get nearly enough attention as misogyny, despite the fact that misogyny and white supremacy are constitutive of each other. They are, she said, "of a piece."

White supremacy is what Alex implicated in her op-ed—the same belief that animated the people who would call her slurs, the same belief she suspected influenced her professor's reaction after the op-ed ran and which she believes also explains why some of the white women in her class did not defend her that day, a silence that tells its own story about white women's complicity in Black women's oppression.

In 2017, shortly after the first inauguration of President Donald Trump, I interviewed Kimberlé Crenshaw, a leading critical race scholar who coined the term *intersectionality* to describe the unique combination of racism and sexism Black women face. I asked if she would characterize the moment and explain what was at stake. I was so naïve that day.

She told me we have acclimated to the violence women face. She said a system of power is so normal that even those who are subject to it are internalizing and reproducing it. Remember that in 2016 nearly half of white women voted for Trump. Never forget that less than 1 percent of Black women did. In 2024, white women helped deliver Trump another win.

After Trump's 2016 election, there was a dramatic uptick in harassment and intimidation targeting people of color, immigrants, and the LGBTQ community. Data from Pew shows that since 2017, online harassment has gotten more severe. A 2018 report from TrollBusters and the International Women's Media Foundation, authored by TrollBusters founder Michelle Ferrier, found that since 2013 online attacks against women journalists have become more visible and coordinated, and a separate global study from the International Center for Journalists authored by Julie Posetti and Nabeelah Shabbir found that Black, Indigenous, Jewish, Arab, and lesbian women journalists experience the highest rates and most severe impacts of violence online. During the 2024 presidential election, *The New York Times* reported that lax social media controls were partly to blame for a spike in xenophobia and hate speech during the race between Trump and Harris. In the days after Harris's loss, the Institute for Strategic Dialogue said there was an onslaught of online abuse and harassment directed toward women.

Black women's experiences of abuse have been historically minimized and sometimes outright erased. Their prescience about the dangers of a nascent alt-right were largely ignored, and at least some of the online harms people experience today are a result of white people, including white women, refusing to heed Black women's warnings. Black women's pain is rarely deemed worthy of serious attention, which was precisely the point Alex made in her op-ed when she denounced the dance department's lack of protection for Black women.

"They completely disregarded my feelings because in their minds, I was not capable of feeling," she wrote.

Words

After Alex's op-ed was published, she tried to ignore the abusive messages people sent her online. She told herself: "These people are just crazy. These people are wild. They're insane. I don't care."

But she did care. There are so many ways that words matter.

In her book *Seven and a Half Lessons About the Brain*, neuroscientist Lisa Feldman Barrett—who collaborated with PEN America for the "self-care" section of its field manual against online harassment—explains that our brain's most important job is to manage our body's metabolic budget so that we can stay alive. We are not conscious of every thought, feeling, kindness, or insult functioning as a deposit or a withdrawal against that budget, but she says that is precisely what is happening inside us. Words that are generous and connective function as deposits, while words that are degrading and exclusionary function as withdrawals.

Words have potent effects because the regions of the brain that process language also control major systems that support our body budgets. These are the parts of our brains that guide our heart rates, adjust glucose, and manage our immune systems. "The power of words is not a metaphor," she writes. Words are "tools for regulating human bodies. Other people's words have a direct effect on your brain activity and your bodily systems, and your words have that same effect on other people. Whether you intend that effect is irrelevant. It's how we're wired."

Unless we can keep language from getting through—which we can sometimes do online with mutes, blocks, filters, and other people screening our inboxes and mentions—the brain has to process language. Negar Fani, a clinical neuropsychologist and associate professor of psychiatry and behavioral sciences at Emory University, says that anytime the brain receives information, it necessarily identifies whether the information is important and whether it is a threat to survival. The brain also has to derive meaning. The power of words lies in their social and political meanings. As linguist Sally McConnell-Ginet writes in her book *Words Matter,* it is our histories and social practices that "pour content into words, endow them with meaning and power."

Knowing the history of a word makes it that much harder to set aside. CNN race reporter Nicquel Terry Ellis and I were former colleagues at *USA Today,* and I interviewed her for this book because I witnessed her abuse, which was differently textured than my own. She was called not just a "cunt" but a "house slave." When we spoke, she told me about the time she was home on the couch, exhausted after a long day, and saw the message that had been posted on her Facebook fan page: "N——."

She thought about her former manager, another Black journalist, a person she respects and admires, who told her she should try to ignore this language, who stressed that Black people faced worse during the civil rights movement—bombings, lynchings, beatings. Ignore it and do the work, he encouraged.

She couldn't. Terry Ellis told me: "To have someone send you a message, someone taking the time out of their day to send you a message and call you the N-word . . . a name that was given to your ancestors who were slaves, who were called that by slave masters when they were told to get out in the field and pick cotton. I mean, you think about all the historical implications of that."

No matter how many cognitive tricks we employ, certain words demand the body's attention. We cannot think our way out of every reaction.

Jennifer M. Gómez, a Black feminist sexual violence researcher and assistant professor at Boston University's School of Social Work, told me the slur was directed at her when she was Zoom bombed—a disruption from harassers during a video conference—during a virtual awards ceremony hosted by the American Psychological Association. The interlopers put up swastikas. They shouted "fuck the n——" over and over. The audience left. The audience tried to return, but it happened again. Gómez left and did not go back. She sat at home, whispered to herself that it was fine, that she was safe in Detroit, that the interlopers were not in Detroit. But she did not feel fine. She wanted to take a walk. Is it safe? she wondered. She cried. She felt anxious. She scolded herself. She was a violence researcher who should know better. How dare she be shocked?

Gómez had not been prepared for abuse during the ceremony. When she marinated on why the words felt so violating, she realized it wasn't that the transgression occurred during a private event; it was, she said, that it "happened within my home."

• • •

The words people use to speak to us and about us tell us a great deal about how other people see us, which impacts how they treat us. Misgendering or deadnaming a trans person doesn't happen in a vacuum. It is connected to laws being passed in legislatures across the country that deny trans people human rights. Calling a woman a "cunt," reducing her from a human to an anatomical part, is connected to a rape culture that makes sexual violence against her permissible. Calling a Black woman a racial slur reinforces her position in a white supremacist society that values white life above all other life, that demonizes Black bodies and brutalizes Black bodies, often without consequence.

I have told my children that it doesn't matter what others think, only what they themselves think. I have asked them: *Do you like how you look? Do you like how you sound? Do you like who you are? Yes? Good.* But recently I've stopped asking these questions. How other people see us, how they speak to us, shapes our lives—the privileges we are afforded, the dignity we are denied. As Gloria Steinem told me the second time I interviewed her, this time over Zoom, in a pandemic-weary world, "Sticks and stones may break my bones, but words can never hurt me" can be a useful defensive tactic, "but it is also true that we are social animals, and we get our feelings about ourselves from other people's perceptions. The perceptions come . . . heavily in words. That's very important."

One night I was swiping through TikTok at an unreasonable hour when the algorithm delivered Jools Rosa (who would later create the viral "demure" trend). Rosa is a trans plus-size Afro-Latina beauty influencer with more than two million followers on the app. In the video, Rosa painted the word *fatty* on her chin in color corrector and blended the insult into her face. Impressed with this symbolic approach, I scrolled through her feed. I saw a woman challenging stereotypical depictions, fiercely funny, serially self-deprecating, and at times painfully vulnerable. On a trip to Las Vegas, Rosa posted videos of herself in full glam to watch Beyoncé perform, posted another at the pool lauding herself for not sweating off her makeup, posted another divulging that she had met a man the night before. She felt good with that man, connected, but she is a trans woman in America, her safety is routinely threatened, and she started to question her reality, to grow paranoid. She convinced herself the man was going to round up people to assault her, so when he went to the bathroom, she slipped away.

I reached out to Rosa, and when we spoke, she told me she is subject to a daily torrent of racist, sexist, transphobic, fatphobic messages online. Content creators with large platforms face a heightened risk of online harassment. Men call Rosa disgusting. Children call her a gorilla. Thin, white, passing transgender women deride her for not being trans enough. One of the strangest parts of the abuse, she said, is how it morphs into a preoccupation with the way not only people on the Internet but everyone sees her. Someone once commented on one of her videos that if they ended up sitting next to her on a plane, it would ruin their entire flight. She carries that now, that specific wondering about what fellow passengers think of her.

"I start picking up on how I perceive people are perceiving me.

I'm like, great. Everyone thinks I'm a nasty bitch. Everyone's looking at how big I am. Everyone's disgusted by me."

Working out how other people perceive us is an important part of understanding communication. It's why a lot of the online abuse you would not think would demand our attention does, especially some of the less obvious kinds. It's easy to assume what a person sending a rape threat or death threat thinks of you. It takes more work to sit with the subtler messages. Linguist Emily Bender told me that understanding language includes imagining what the other person is trying to say.

"Even if we are able to set it aside afterward, we still have to have made sense of it, and it's very, very difficult to do that sense-making without modeling the mind of the person who said the thing," she said. "It's intimate."

Emotion Regulation

I don't want to suggest violence should be a woman's problem to solve. I don't want to suggest that there is a single solution, neatly wrapped, that she can take into her life, into her work, into her body to feel immediately better or stronger or more resolute. I won't suggest that there is a way we can feel better about sexism or racism. I do not want us to feel better about sexism or racism.

I began this book wondering how women were surviving violence online. I wanted to know how women coped. I thought there would be a clear number of beneficial strategies and I would be able to sensibly arrange them.

I have learned a great deal about how women cope, but I cannot be honest about those findings and package them as I'd hoped. Coping involves a number of strategies influenced by alternating priorities. Everywhere there are binds, and everywhere women are getting tied up.

What Alex knows with certainty, what she told me again and again, was that she tried to ignore the abusive messages but could not. She told herself she was unbothered. Her instinct to ignore her feelings is part of the unseen labor that Black women and other people who experience oppression perform daily. As feminist scholar Sara Ahmed writes in *Living a Feminist Life,* sometimes "surviving the relentlessness of sexism as well as racism might require that you shrug it off, by not naming it, or even by learning not to experience those actions as violations of your own body; learning to expect that violence as just part of ordinary life; making that fatalism your fate."

Like Alex, many women online try to regulate their emotions, to control what they feel and express. Emotion regulation is defined by psychologist James Gross, a pioneer in emotion research, as "the processes by which individuals influence which emotions they have, when they have them, and how they experience and express these emotions." Emotion regulation can be done unconsciously, and it can be taught. The strategies people use to regulate their mood and behavior can be based on what they believe about emotion, and their tactics are also influenced by what is culturally accepted about emotion.

When I researched emotion regulation, I noticed two common strategies in the literature: expressive suppression, when a person tries to hide an emotion, and cognitive reappraisal, when a person tries to think differently about a situation to change an emotional response.

If you are engaged in expressive suppression, you try not to let anyone around you see how you really feel. I think of Alex in her dance class, the way she controlled the muscles in her face so no one could read her distress. After the response to her op-ed, Alex told me, she did not want people to know how angry she was. She feared playing into the stereotype of the angry Black woman. Black adults are more likely to suppress their emotions than white ones. Suppression can be physiologically and cognitively taxing and can decrease positive emotions.

Cognitive reappraisal is generally considered a healthier emotion regulation strategy. Sometimes it is used to dull a negative feeling or to feel something else altogether. It may involve going over a situation in your mind several times to come up with an alternate take. You might see the situation one way, then try a broader perspective. Maybe you attempt to find the silver lining.

If you go to a restaurant and the waiter is rude to you, you might ruminate on how awful that felt or how undeserving you are of that mistreatment, or you can attempt to reappraise the situation to consider that that person might have been having an awful day and that their impoliteness may not be about you at all. In the context of online abuse, reappraising a situation can look like changing your thinking from "people are saying horrific things to me online because I said something wrong or did not make my point in the correct way" to "people are saying horrific things to me online because I said something that needed to be said, and that is threatening."

Reflecting on my own abuse, I believe at the start I heavily suppressed my emotions. I didn't understand why I couldn't ignore my abuse, as so many other women seemed to be doing that. (I now view this as a misunderstanding, a symptom of our legitimate reticence to disclose just how difficult these experiences are.) I also ruminated, which sometimes kept me preoccupied with my own experiences in ways that prevented me from more broadly analyzing how my story fit within a larger one. I likely also used reappraisal as a way of meaning-making, which I believe was helpful, even when fraught. I often evaluated abusive messages mention by mention. The nature of journalism demands a degree of humility, a recognition that it is always possible we may get it wrong. Whatever story I told myself about a piece of hate would rarely stick, because I found myself appraising hateful responses on a case-by-case basis. It took a long time for any part of the story I told myself about what I did not deserve to become self-sustaining.

When I began this research, I thought emotion regulation

sounded like a reasonable strategy for coping with negative feelings, especially the tactic of reappraisal, which the literature suggests is positive for well-being. Things feel awful, things are awful, so you do what you can to manage your emotions, to feel better. We cannot snap our fingers and rid the world of violence, so we might as well learn to regulate our emotions around the violence we face, minimizing their disruptions. We can't ignore the abuse, but we can reframe a situation to dismiss individual instances of hate and maintain a sense of self-worth, to continue to participate in public life. Discrimination is terrible for health. It makes people more anxious and depressed, and it prematurely ages the body. If we can manage our emotional responses, we can stay healthier in the face of oppression. If we are healthier, we are better equipped to fight oppression.

But when I dug deeper, I began to find the trouble.

In 2018 Alfred Archer, a philosopher at Tilburg University in the Netherlands, read a paper by philosopher Amia Srinivasan in which she writes that victims of oppression are often asked to turn away from valid emotional responses to give themselves the best chance of ending that oppression. This choice, argued Srinivasan, constitutes an affective injustice, since it forces victims of injustice to make a tragic choice between emotionally recognizing their experiences and doing what they can to address them. Archer was intrigued by the paper and was contemplating its arguments when he attended a conference where the renowned emotion researcher James Gross gave the keynote speech. Archer and fellow Tilburg philosopher Georgina Mills began to think about the relationship between affective injustice and emotion regulation, eventually concluding that "the demand faced by victims of oppression . . . is a demand that they regulate their emotions."

Srinivasan's paper explored the aptness of anger, so Archer and Mills examined emotion regulation strategies in the context of anger, too, though their argument could apply to many other emotions. Archer and Mills looked at the strategy of cognitive reappraisal, which they said may make a woman feel emotionally better but can also involve "turning away from the injustice." They also looked at expressive suppression, which they note may allow a woman to remain clear on an injustice but can lead to "harmful outcomes."

To illustrate the problem with reappraisal, they offered the example of a woman who is angry at being sexually harassed, cognitively reappraises the situation to decide it wasn't harassment, and then essentially ignores the problem and "gives up on attempts to challenge the injustice." (The goal of reappraisal is to change your emotion, so the story you settle on may be one that makes you feel better but not necessarily one that is true.)

To illustrate the problem with suppression, they give the example of a woman who does not express her anger. That choice may keep her safe from further discrimination and avoid making other people around her uncomfortable, but it can decrease her positive emotions, cause problems with cognition, and lead to poor health outcomes.

When Archer and I Zoomed to lean into these complexities, he admitted that while some form of emotion regulation is necessary to survive life, in the context of oppression, the act of trying to control your emotional state is fundamentally unjust.

"The fact that you have to try and work out how to respond, how to keep yourself healthy in the face of this, shows how big a difficulty it is. Because not only are you faced with all of these options, each of which brings costs, but thinking through how you are going to respond to a situation is enormously cognitively taxing in itself."

There are other complications to consider. A 2019 paper led by psychologist Brett Ford at the University of Toronto found that reappraisal is not always healthy when a person reduces their negative emotions in a way that leaves them feeling inauthentic and it is not always effective when the situation causing them discomfort is especially intense. Most emotion regulation research is not done in the context of discrimination or identity-based attacks. There's a difference, of course, between a rude waiter and a Nazi in your mentions.

When researchers do look at emotion in the context of discrimination, they reach more nuanced conclusions. A 2022 study by scholars Ajua Duker, Dorainne J. Green, Ivuoma N. Onyeador, and Jennifer A. Richeson looked at how three emotion regulation strategies affected outcomes in women who had experienced sexism: self-immersion, which typically involves rumination; self-distanced reappraisal, which involves taking a third-person or bird's-eye view of an event; and positive reappraisal, which involves finding a silver lining or the positive outcome of an experience. When participants used self-immersion, researchers instructed them to ask themselves questions like "Why did I feel this way? What are the underlying causes and reasons for my feelings? How did I respond?" Those using self-distanced appraisal were instructed to ask themselves those same questions using a third-person perspective, so instead of "I," they used their name: "Why did Jane feel this way? What were the underlying causes and reasons for Jane's feelings? How did Jane respond?" Participants who used positive appraisal, and specifically a redemptive narrative about overcoming adversity, were instructed to reflect on lessons they learned, how they grew, and how the experiences shaped how they felt and thought and even who they are.

While a strong body of research suggests that self-distanced cognitive reappraisal can relieve psychological stress, it did not make the participants in this study feel any better than rumination. It was the women who reflected on an experience of sexism using positive reappraisal with a redemptive narrative that had the best outcomes. The study showed that in the context of discrimination, it is not enough to suggest a woman use cognitive reappraisal to feel better. It matters how she uses it, in what specific ways she reflects on a situation, and how she integrates the experience into her identity and life.

José Soto, a professor of psychology at Penn State who has conducted research on how culture can influence emotion regulation, has also found distinctions around discrimination, showing that reappraisal is less helpful in the context of oppression for Latinos, and that while expressive suppression is linked to poor psychological functioning for European Americans, it is not for members of East Asian cultures.

Soto said that when a person is trying to manage emotions around aggressions that are group based and identity based, the effectiveness of emotion regulation strategies is likely going to be influenced by that person's goals.

"Any one person in any one situation that's characterized by discrimination or oppression can have a different goal," he told me, "and I think we have to also keep that in mind. So some person might be thinking, 'I'm just trying to preserve myself. And for me to do that, I have to do X, Y, and Z.' But for somebody else, their goal is to fight the oppressor and let them know that this is not okay. And so those different goals will lead to different approaches."

Alex's goal was to survive her senior year. And she did. She is a doctoral student now, studying how the messages of fascist groups impact the psychological development of Black girls. She organizes in Boston, writes for the Global Network on Extremism and Technology, and conducts research on far-right misogynoir.

"It doesn't stop," she told me of the abuse. "Anytime I write something, it doesn't really stop."

Alex told me the difference now is that she doesn't "pretend it's not there."

Learning how to regulate emotions is important for psychological well-being, but the strategies we use can force us into near-impossible choices about what is good for the moment and what is good for the future, what is good for the individual and what is good for the collective. I struggled with how to conclude this chapter, because I wanted to end with an idea that felt unequivocal and concrete. That urge was impossible to satisfy, so I landed on a conversation with psychotherapist Seth Gillihan, who teaches people how to regulate emotions by changing unhealthy patterns of thinking. Gillihan, who had been a source during my reporting on the trauma of the January 6 insurrection, told me that if someone is upset about their abuse and is beating themself up for being upset, wishing they had thicker skin, thinking they can't do this work or that they shouldn't be sharing ideas publicly if they can't handle them emotionally, he would encourage them to notice those thoughts, to loosen those attachments, and to question those assumptions.

"Maybe you could ask yourself something like 'Do any of the people I know who have expressed who they truly are in spite of society's criticism or hatred, who have really changed society, have they done it without some level of abuse? Is it worth what it's going to cost?' And the answer might be no. But someone might realize, 'Oh, nothing says this should be easy.' And there can be real relief in that. And realizing this is hard. Yeah, it's hard. Exactly. That's exactly how it is."

Pain and reprieve are the nature of struggle. They formed the poles of Alex's experiences. They made their way into her art. When I asked Alex for some examples of her choreography, she shared a piece she created during college that featured five Black alums. The piece began with the women moving into a circle, clasping and raising their hands before breaking into smaller groups. One performed a solo, then they came back together, eventually dancing in unison. You don't have to understand every move to feel the weight of it. Alex told me the piece was about how Black women are rarely protected, so they depend on the love between them to invent safety, to sustain movement.

Alex told me she later watched a recording of the performance. When it ended, in that liminal silence before the audience claps, you can hear a single voice, low and proud: "Yes."

It was her mother.

CHAPTER 2

Words Will ~~Never~~ Hurt Me

COMEDIAN MARIA DECOTIS is out with a friend who works for a movie producer. The producer is a notable name, and he made an acclaimed movie, and there is a big premiere. She and her friend go to the premiere, to the after-party, and then to the bar. They are drinking and talking and laughing. She looks at her phone, opens a message request on Instagram. She always opens them, even when she doesn't know the sender, because it could be a producer or a director asking her to audition. She can't ignore opportunities.

The message is a picture of a naked man. He calls her a "dumb bitch." He tells her she's "not fucking funny." She wonders if he knows where she is. She gets recognized sometimes now, on the subway. She says it creates a sense of always being watched.

Maria shows the messages to her friend, who is a man. He is flabbergasted. She is flabbergasted, too, that he is flabbergasted. "I get these all the time," she tells him.

Her abuser's name is Big Kyle, "magakyle" followed by a trio of digits. She makes a joke about his name. They laugh, but then the joke is gone, and the messages keep coming. He goes on and on. Terrible things, humiliating things.

She posts his messages on Instagram Stories in real time, so people can see. How dare he, she thinks. She can't stomach the audacity. And if he is near, watching or waiting, and something happens to her, then, she thinks, people can see and tell someone that maybe big magakyle had something to do with it.

She is angry that every time she posts a stand-up clip on social media, some guy comments on her face. She wants to post the messages on Twitter, but she is afraid. Things can blow up on Twitter and get chaotic. She is worried that with a less intimate audience than Instagram, people will blame her. She worries they would ask: What did you do to upset him? Why didn't you immediately report him? Why are you posting this? (*What were you wearing? What were you drinking? What did you do?*)

She thinks back to a decade ago, when she was twenty-one, visiting her aunt in Italy. One afternoon she went out alone, and a man on his motorino stopped and said, "Come home with me. Will you have sex with me? I'll pay twenty dollars if you have sex with me."

It was a small town, and by the time she got back to her aunt's house, word of the man's proposition had already reached her.

Why were you out there? her aunt asked.

I was taking a walk, Maria said.

You need to stay home now.

Two months after magakyle sends his messages, Maria works up the nerve to post them to Twitter for everyone to see. Her only edit is a strategically placed emoji over the most offending parts.

I reach out to Maria after I see her posts. When we speak on the phone, neither of us laughs.

"This experience is now inside of me," she tells me. "It's now hurt me. It's now fucked up my brain. It's now set me back on where I was feeling in terms of my work and my creative process."

Maria and I talk about the ways women fight for dignity and safety, the struggles that are easy to see and the ones that others rarely can. She tells me: "This emotional and psychological labor that we're doing just to keep up with all of this, to stay ahead of it, to try not to let it get to us, is a whole other battle that we're fighting in our brain. Every day this is what we're up against, is people telling us to disappear. They want us to disappear. . . . They want to poison us until we take it upon ourselves to disappear."

After she posts the screenshots, magakyle tells her he cannot have his "big dick" online like that. He warns her, "This better not go viral bitch."

It goes viral.

Sexual Violence

Maria experienced sexual violence online. When we spoke, she called it violence. She described the violence of language and the violence of a man heaving his genitals into her DMs. We can start with the sexual violence, with the most obvious violence, with the parts of her story that make us cringe or gasp, that made her sick and afraid. But we won't stay there, because this isn't just about dick pics, about the worst things men say, the most obscene things men do. Maria wasn't just disgusted by magakyle's genitals. She wasn't just afraid. She may have been ambiguously fearful that magakyle would find her and physically assault her, but she was certain she was emotionally hurt. The experience was socially painful.

For Maria, these messages topped a pile of others that made her question her talent and her belonging as a comedian. When we spoke, she didn't stress her fear, she stressed the slow annihilation of a self. This is why we can't focus just on what the abuse looks and sounds like—we need to look carefully at what the abuse does. Sometimes it can rattle the most vulnerable parts of who we are. When we pay attention only to the most egregious features of online abuse, we misunderstand its power, how insidious it can be. We'll start with the sexual violence, but we'll follow the pain. When we follow the pain, we find a truer story.

The sexually violent nature of women's online abuse is perhaps its most distinct feature. We are told we deserve to be raped, we listen as people describe how they will rape us. Men, in particular, use the sexualized language of sluts and whores whenever women try to speak or perform or participate in public life, whenever women enter spaces where men do not think we belong. Women are more likely than men to report having been sexually harassed online, and young women are particularly vulnerable.

Jennifer Scarduzio is an associate professor of interpersonal and organizational communication at the University of Kentucky who studies emotion, violence, and identity. She said that her research on female college students who were sexually harassed across multiple settings shows how the Internet adds complexity to the experience. Sexual harassment online is normalized, and because it is happening on our phones, it is taken less seriously by everyone from friends and family to law enforcement. But it is precisely because it is happening on our phones that it is maddeningly difficult to circumvent. If you're being sexually harassed in person by your boss, you can still leave work at the end of each day. If you're being harassed in person by another student at your school, you can leave common campus spaces and return to your dorm. But when, like Maria, your sexual harassment is in your DMs that you regularly check, when it is mixed in with legitimate work opportunities, it becomes nearly impossible to avoid.

Sexual violence is a pervasive social problem, and these experiences online remind us that men will use sexual violence to exert power over us anywhere we are. Over half of U.S. women have experienced sexual violence involving physical contact during their lifetimes, and even this number underestimates the scope of the problem, as many women are too ashamed to report it.

Maria spent months contemplating whether she should post magakyle's messages to Twitter. These abusive messages are designed to keep us quiet, to make us feel so ashamed that we feel we have no choice but to remain silent. Many women are embarrassed to be associated with lewd sexual language. Who wants to tell their male boss that someone emailed them about "guzzling it down" and "slurping it up"? Who wants to show people who care for them how badly they are spoken to? Maria's parents follow her on Twitter.

Magakyle's dick pics capture our attention. This violence is worth our attention. But when Maria spoke with me, she was not overly preoccupied with dick pics. Instead, she was focused on how a stream of abusive messages like magakyle's had created a self-consciousness about her comedy that was far more pernicious than the routine self-reflection all artists undertake. The pain Maria felt tied to the fear that maybe, actually, she wasn't fucking funny.

Maria doesn't believe men who harass her online desire her sexually. She believes they want something else entirely. She believes they are saying: "I want control over you, and I want sexual control, and I want to tell you what to do, and I want you to shut up. I want you to shut the fuck up."

Some women attempt to hide their identities online to avoid sexual harassment. I think about the female gamers who use gender-neutral pseudonyms and avatars. Some avoid speaking to other players, because the sound of a woman's voice is enough to incite harassment. It does not matter what she says, only that she is speaking.

But concealing gender is not an option for women with public-facing professions, women, like Maria, who are trying to build careers that require a degree of recognition, fans, an audience. Some of us cannot or do not want to pretend that we are anything less than who we are.

Hurt

The violence may be ubiquitous, but the suffering is specific. What hurt Maria was the way these messages threatened her self-concept, the positive beliefs she holds about herself, or tries to. Violence online doesn't just threaten women's participation in public life, it threatens the construction and maintenance of a woman's sense of self, which is in part a reflection of the reactions of other people.

Online abuse is a problem that goes beyond the hurt of individual women, but the hurt that individual women experience matters.

I have been hurt by things people have said to me online—when they told me I was terrible at my job, when they said I was a bad mother, when it was suggested the makeup of my body disqualified me from speech. I suspect that at least some of this hurt is felt because of what patriarchy has given me, a sort of mental tentativeness, a physical self-consciousness. Sometimes it can be hard enough to look at yourself in the mirror. A hall of mirrors where you are constantly getting feedback that you weren't asking for is not an environment I could ever well tolerate.

Emotions are not frivolous. When we experience and express emotion around our abuse, we are not being oversensitive or overwrought. Emotions help us make sense of ourselves, other people, and the wider world. They give us information that helps us decide how best to behave, whether to move toward something or away from it, whether to shout or run, and sometimes whether it is safe to speak. When people communicate with us in ways that are hateful, demeaning, or degrading, when they use language to reject us, disconnect us, or isolate us, reacting emotionally is natural and useful. When someone suggests you ignore abusive messages online, when they tell you not to let it hurt your feelings, they're asking you to deny responses to communication that are there to keep you alive.

Emotions are also social and political. Beliefs about emotion are linked to beliefs about gender, race, class, and culture. Scholar Stephanie Shields, who studies the intersections of psychology, gender, and emotion, wrote in a 2005 paper that the politics of emotion dictate whose emotion is valued and whose emotions are dismissed as overreaction. Historically, she writes, when emotions are associated with men, they are thought of as being in the "service of reason," while those associated with women are portrayed as "inferior" and "ineffectual."

Several years ago I attended a women's journalism conference in

New York, and an older woman panelist fielded a question from a young reporter on whether it was okay to cry at work. As she began to answer the reporter's question, my mind flashed to the many times I had cried at work: When I combed through photographs of bodies ravaged by a chemical attack in Syria. When I wrote of the trauma of Palestinian children after watching a girl in Gaza gesture to the ruins around her, pleading, "What do you expect me to do? Fix it? I'm only ten." When the schoolchildren were gunned down in Sandy Hook, when I tried to make sentences out of the dissonance between their bright smiles and the facts of their death. When I interviewed a mother whose son had died by suicide and then sobbed in an editor's arms.

The panelist told the audience that women should not cry at work. She worried that men would use it against us. I do not want to appear obtuse about the challenges women face. I am aware that perceptions of our competence can be influenced by perceptions of our emotionality. But her comment did not sit right. I want to be moved by tragedies, to empathize with victims. I do not want to uphold stoicism as a virtue, as a model. Don't men want to cry about the dead children, too?

Women are entitled to our emotions. We are allowed to take them seriously. Radical feminist Kathie Sarachild wrote that when detractors of the second-wave consciousness-raising groups referred to their gatherings as "bitch sessions," the women would respond by saying, "Yes, bitch, sisters, bitch."

For Maria, the messages from magakyle emphasized the daily drumbeat of "you're not funny," which hurt her because it touched a major part of her identity. In a chapter for the *APA Handbook of the Psychology of Women,* Stephanie Shields wrote, "The central feature of emotion is that it concerns an event, situation, or object that the individual perceives as potentially having significance for her or his personal well-being." When we say we are emotionally "hurt," it often means that something has penetrated the self, has called into question something about who we are. Anita Vangelisti, a professor of communication at the University of Texas at Austin who has spent years researching the concept of hurt, said hurt is distinct from other emotions, but it is often experienced with other emotions, including anger, sadness, and humiliation. Hurt's defining feature, she told me, is that it "involves vulnerability."

When Maria and I are on the phone, I can hear the hurt in her inflection, in her pauses, in the questions she asks herself.

"Every comedian, every artist has to have an encounter with themselves over and over again about the floor falling out underneath them, whether or not they should really be doing this. 'Am I really an artist? Am I good enough? Am I doing the right thing with my life? Have I wasted my life dedicating it to something that doesn't matter?' Every artist goes through that sort of transformational death every once in a while, just simply as being an artist. But then on top of that, you have someone begging you to do these deaths, begging you to go away, begging you to shut up. Demanding, not begging, demanding that you shut up and stop what you're doing because you are not good enough."

Maria is candid with me about her hurt, but she's not broadcasting her pain. Online, vulnerability can be a snare. Many women don't want to be accused of complaining too much. Pain is blotted out. Lyz Lenz, an Iowa-based writer and author who publishes the irreverently titled newsletter *Men Yell at Me,* told me that online she generally projects an unaffected persona, which has sometimes confused people in her life when she tries to confide in them about her abuse.

"I remember talking to a friend about how upset I was and him being like, 'Oh yeah, but you're Lyz. You'll be fine.' And I was like, 'Oh, he sees me as this persona that I put on, but he doesn't see me.'"

Lenz has experienced a torrent of online abuse over her coverage of politics in red America. She has been targeted for her response to a crude Bernie Sanders supporter during the 2016 Iowa caucuses, for her profile of Tucker Carlson, and for her reporting on accusations of domestic violence against alt-right figure Richard Spencer. Her abuse has included threatening tweets and emails, bomb threats, Bible verses mailed to her home, claims that people were going to call CPS to report her for being an unfit mother, and images of Pepe the Frog raping her.

When we speak, Lenz is Midwestern wry, deft at jokes that blend so seamlessly into conversation that it can take a moment to catch the quip. She told me that at the worst of her abuse, she felt as if she had lost herself, had been destroyed "by the dumbest, most boring people in America," which she found offensive. "There are a lot of different reasons it hurts, but also it could just be a bad day," she says. "Your car could have a flat tire, your ex could say something shitty, some dude ghosted you, your kid said you were ugly. And then you open up your email and some condescending old man is telling you that you don't know what you're talking about. And then you just cry."

There are so many cleverly malicious ways to cut through to the deepest parts of who we are. One of the worst comments that disinformation researcher Abbie Richards told me she ever received was from a person who said she was having "a net negative impact on the world."

She told me: "They find a way to get at the things that mean the most to you."

Social Pain

Experiences of online abuse can also cause pain when they function as rejection and disconnection. These experiences can make us feel as though we don't belong—on the Internet, in comedy, in journalism, in technology, in academia, in politics, in public life. Humans are social creatures who want to belong. When many of us say our abuse hurts, we are describing not necessarily our distress around our self-concept but our experience of social pain, which involves threats or damage to a person's sense of social connection or value.

UCLA social psychologist Naomi Eisenberger told me experiences of rejection or exclusion activate some of the same parts of the brain that respond to physical pain. Evidence suggests that in the body there is an overlap between the physical pain system and what she calls the "social pain system." We can feel social pain when we aren't picked for the team, when someone breaks up with us, when a person uses language to insult us. And because our bodies perceive rejection as a threat to safety, those rejections are painful even when they come from members of a group to which we do not intellectually want to belong. They can especially hurt when we are rejected by a group to which we do.

Many of the women I interviewed said that some of the most difficult and painful experiences of online harassment involved people who shared some aspect of their identity. Communications scholar Alice Marwick has found in her research that people across the political and ideological spectrum engage in abuse online as a way to enforce social norms. They use moral outrage to justify harassment, which is at least part of the reason why you also see progressives attacking other progressives, feminists attacking other feminists, Black women attacking other Black women. Sometimes we silence one another.

I could see Marwick's explanatory model in disability activist Kara Ayers's story. Ayers told me the most painful experiences of harassment online were perpetrated by members of her own osteogenesis imperfecta (OI) disability community after she committed what some members perceived to be a moral violation. Osteogenesis imperfecta is a genetic condition that causes bones to break easily. Ayers has biological children, and she found herself viciously targeted as her community weighed in on the ethics of her decision. With each of her pregnancies, her child stood a chance of inheriting the gene. Neither of her daughters was born with OI, but the potential they could have been enraged some members of her community.

"The most hateful things have actually come from other people with OI in terms of their belief that it's a life of pain and suffering and that I, of all people, should know that," Ayers said.

Ayers doesn't believe the world would be better off without disability. She experiences pain, but she told me that doesn't mean she wishes she were someone else: "It's not like my life is painful and so therefore it's not worth being who I am."

The online abuse itself functions as social rejection, but so can community responses after women disclose. Some of the most distressing experiences around our harassment are felt when we reach out to members of our community—friends, family, co-workers, employers—and they minimize our abuse and suggest we are not thinking about it in the right way or are not coping with it in the right way. It can hurt to hear them say, "You decided to post online, what did you expect?" Or "Why are you responding?" Or "Why have you not logged off?"

Instead of asking, "Why didn't she log off?" which can sound dangerously close to blaming the victim, perhaps we might ask, "Why should she have to?"

Humor

Maria's story started with a joke. We can return there. As a comedian, Maria uses humor to cope with her experiences of violence and to fuel the narrative engine of her art. Many men suggest women and especially feminist women are humorless, that we don't make good comedians. But they don't understand how much laughing we have to do just to survive.

In the basement of a Brooklyn bar on a sticky July night, Maria stood illuminated before a crowd—Gen Z women, millennial women, straight men, gay men, a boomer couple making out. Victims' rights lawyer Carrie Goldberg was in the audience. Maria couldn't see us, but we could see her—giant eyes, wild hair. She was effervescent:

> I'm actually such a good comedian, when I tell people I'm a comedian, they laugh. Yeah, guys will try to hit on me and they'll be like, "So what do you do?" And I'm like, "I'm a comedian." And they're like, "That's so funny because you look like you make tiny cupcakes for a living." . . . But these are actual real reactions that I've gotten from men when I tell them I'm a comedian. . . . For some reason it becomes a police interrogation. They're like, "At what time did you start comedy? What drugs are you on, crazy girl? Don't lie to me. Just tell me you're not a comedian and I'll let you go home. But if you continue to lie to me, you're going to comedy court. The judge is Louis C.K. He's innocent, by the way."

Like Maria, many of the women I interviewed for this book used humor to mitigate psychological and social pain. They made jokes to me. They told me about the jokes they made to others. The humor makes its way into the work, like Lyz Lenz's newsletter, which features a "Dingus of the Week."

As philosopher Cynthia Willett and historian Julie Willett write in their book *Uproarious,* humor "can serve as a source of empowerment, a strategy for outrage and truth telling, a counter to fear, a source of joy and friendship, a cathartic treatment against unmerited shame, and even a means of empathetic connection and alliance."

Laughter feels good. It can act as a buffer against stress and help us regulate emotions that improve psychological well-being. It has physical health benefits, and research shows it can reduce experiences of physical pain. Humor is a means of fostering social connection, as jokes are often made in the company of others. Humor can help us dismiss some forms of emotional pain and even derive meaning from it.

Disinformation researcher Abbie Richards, who has a background in stand-up comedy, told me when she highlights the absurdity of something, it lessens its power. Richards said one time a conspiracy theorist put up a fake article on Academia.edu about her being in the deep state, claiming the paper explored "the deleterious effects and dark empathic personality of a POS simpreme."

Because it was an "academic paper," the author gave himself a real grade: B minus.

"The only thing was to joke about it," Richards told me. "He wrote nine pages on me and then decided it was worth a B minus." She said that when she joked about the "academic paper," it shrank her abuser down. It helped with dismissing the abuse, because when he was that small, she knew she couldn't take him seriously. It is "a means of sense-making," she said.

I have always used humor to cope with difficulty. I've used it to deflect. Sometimes I've used it to take the edge off fear. Other times to make people more comfortable when they have to witness my pain. Sometimes it feels like if I can make a joke, there is still some part of me that hasn't been touched, some unassailable part that is capable of being clever even under the most painful conditions.

Humor has been effective for me. But there is a limit to what humor can do. In May 2018 I received my first graphic piece of hate when a man on Facebook asked: "Alia, not to get too personal here, but just how much diseased worthless douchebag pussyhound fuckboy cum have you hit your slave knees to suck, fuck and slurp up, eh?"

I felt sick. It was a short bit of a long note. "Pussyhound" four times. "Cum," five. References to "dead toronto whores" (in 2018 a man plowed down several Toronto pedestrians with his van as revenge for sexual rejection) and "schneiderman's whores." (Eric Schneiderman was the New York attorney general who resigned after being accused of abusing women.)

I read it again. It bothered me that it was on Facebook. What parts of me were there, I wondered, that they could see?

I felt like I wanted to make a joke. I imagine I made several, even if I can't remember most of them.

I told my editor about the comment. I know I made a joke to her. I reported it to Facebook.

"The comment has been removed for violating our policies. Thank you for bringing this to our attention," a representative wrote.

I'm certain I made a joke to myself about that. Was this a surprise to them? Where, exactly, had their attention been?

Later that day I called my best friend and made a joke, something about a pussyhound. I can't remember if either of us laughed.

In bed that night I began to read it out loud to my husband. He laughed, but I wasn't making a joke.

I whipped toward him so quickly it felt like my neck might snap. "What's funny?" I asked accusingly.

He looked confused. "That level of vitriol was so insane to me that I thought it couldn't be real. I'm so sorry."

Our union has always been one that reveled in absurdities. But this anecdote encapsulates so well the different worlds that men and women inhabit. My husband doesn't carry around with him the same weight of violent possibility. He heard this comment and thought it was a prank. I read this comment and believed it was a warning.

Where is the humor in that?

The utility of humor is context dependent. Who uses it, how they use it, what it is used to expose, what it can be used to conceal. Sometimes it is okay not to laugh. Feminist scholar Sara Ahmed defines the feminist killjoy as the woman who stands in the face of humor that revels in misogyny. The feminist killjoy sucks the air out of the room. She does not laugh at the expense of herself or others. To be a feminist killjoy is to know when not to engage with humor, when to refuse it.

• • •

While researching humor and power, I came across a 2013 essay by Amy Marvin,[*] then a philosophy student at the University of Oregon and now a Louise M. Olmsted Fellow in Ethics at Lafayette College, in which she looked at how women could, even in an overwhelmingly sexist and misogynist space like Reddit, create their own counterpublics to work against sexist speech. She investigated a group called Shit Reddit Says, where users shared oppressive comments to mock them and discuss their effects. In response to a joke where a Reddit user referred to a picture of two passed-out women as "Jackpot," a member of the Shit Reddit Says group wrote, "Haha I like uncritically repeating things I heard on Family Guy haha also Louis C.K., South Park and Chris Rock haha."

Marvin said that humor can be an effective form of feminist resistance. She gave the example of a London billboard in the 1970s that advertised a car using the tagline "If it were a lady, it would get its bottom pinched." In response, a feminist graffiti artist wrote: "If this lady was a car she'd run you down."

I laughed at that. It felt good. But feminist humor doesn't just offer temporary reprieve, it can also expose, disrupt, and undermine.

In August 2018 comedian Kelsey Caine, a survivor of sexual assault, posted a video on Twitter debuting her satirical character "Penis C.K." She developed it post-#MeToo, after five women told *The New York Times* that comedian Louis C.K. masturbated in front of them or tried to. Most did not consent, some were paralyzed by his ask, and one said she felt pressure to acquiesce. None of the women found this funny.

In the video, Caine pretends to masturbate with a plastic dildo while walking around New York City. She said the character, which she would go on to perform at comedy clubs, was born of frustration. She was tired of watching people, especially other comedians, minimize C.K.'s behavior.

---

[*] Marvin published the essay under the name Amy Billingsley.

"The community wanted to act like they didn't understand why him masturbating in front of his co-workers at their job was a big deal. So I decided I would do that and let them tell me if they thought it was a big deal. And people very much thought it was a big deal when I did it.

"I would say I received a lot more pushback for pretending to masturbate onstage than Louis C.K. did for actually cornering women and masturbating in front of them in the workplace."

Online, she said, people slut-shamed her. They said they hated her. They said no one should listen to her. They suggested she should be "put down." One person said they "would rather listen to my mother beg for mercy while somebody kills her than listen to this." Another called her "brutally unfunny," and remarked, "Women in comedy. . . . Yeah, just don't."

Caine said the only critiques she took seriously were made by other survivors of sexual violence, some of whom told her they found the character triggering. To them, she individually responded.

For the next year, Caine did stand-up as Penis C.K., focusing on New York City venues that were the "most boys'-club places to perform," she said. She said the clubs would book her as Kelsey Caine, and she would show up and perform as Penis C.K.

"I was aiming it at a very specific audience," she said. "I was not trying to preach to my own crowd. I was trying to trigger sex offenders. I was actively trying to upset people who had committed sexual assault, and I did."

When she started performing the character as stand-up, some male comedians with large social media followings attacked her, and their fans took note, inundating Caine with abusive comments. During her shows, many men would make noise, huffing and puffing and stomping around their seats, performing scenes of disapproval. Many men ended up leaving the room. Apparently, they could not take a joke.

• • •

Maria told me her comedy has always been feminist. She remembered the first stand-up material she wrote, scribbled in a notebook, expounding on the ridiculous trend in women's magazines of categorizing women's bodies into different shapes of fruit. Maria said she didn't think her comedy could ever not be feminist.

"It's not something I can subtract from my perspective," she told me.

On Maria's Twitter feed, feminism is threaded through musings and jokes.

Maria tweeted: "You want me to get MARRIED?! The thing that kills 50,000 women a year . . ."

Maria tweeted: "I genuinely think one of the main reasons women haven't overthrown the patriarchy yet is because we've wasted too much energy holding in our farts this whole time."

Maria tweeted: "They want to rape us and then kill us for not giving birth to their babies Happy Women's History Month."

Maria processes pain through creative expression. But there are some things even Maria can't joke about. A couple of days after she posted her thread about magakyle, she said, Twitter removed the post and accused her of violating its policy.

"Twitter said, 'You posted someone's property without their consent,' or something," she told me. "And I was like, 'Consent? Oh, is this about consent now? Because I didn't consent to receive any of this.' "

Maria sees men getting control over the world through violence. Maria attempts to exert control in the world through comedy, so she stands on a stage, an authority on her life, and finds ways to laugh about how absurd it is that women are expected to live as they do.

> We blame white male violence on mental illness, and that's not fair, you guys, because I have a mental illness. . . . We let

> men have a monopoly on violence. And that's not right because women have those tendencies too. I can think of ten men I want to kill right now, and that's my short list. And every day I'm not killing them. I wake up in the morning with my mental illness and I put on my panties one leg at a time, and I sit down and I eat my spaghetti breakfast just like everybody else. Women are out here every damn day doing the work, eating spaghetti for breakfast, and not killing men. And no one's rewarding us.

The basement was alive. Laughter beat the walls. As she walked offstage, I saw Maria's phone in her back pocket.

CHAPTER 3

Where Bodies Live

LEAH TORRES WAS a curious child who delighted in science. She was drawn to biology, to the study of bodies and genes, to the questions of what we inherit and what we pass down. By the time she reached high school, she decided she would become a doctor.

Leah headed to college at the University of Michigan. It was the late 1990s, and while she was there, she learned about a study on uterine cancer,* published in a renowned journal, where the participants were all men. None of the participants had a uterus.

That settled it. Leah decided she would become an ob-gyn, a doctor of women's bodies and women's futures, a doctor who would work to close an egregious research gap.

Leah was doing her residency in Philadelphia at the Albert Einstein Medical Center when her partner suggested she use Twitter as an education tool. Leah loved the idea and decided she'd use the platform to post facts about reproductive health. She shared information but was also compelled to correct lies—the things she saw or read that were not grounded in science, that framed women's health in harmful or disingenuous ways.

In 2013, former NRA spokesperson Dana Loesch tweeted:

---

*Leah Torres does not remember the study, but refer to page 256 of the notes section for a journal article on research gaps in women's health care.

"Every gun ownership doesn't = dead infants. Always getting an abortion = dead infants. #AbortionControlNow."

Knowing that without safe and legal abortion, maternal mortality rates would rise, Leah responded: "Do dead women from forced pregnancy matter?"

That was the first time the mob came.

She remembers how it started. She was in the on-call room at the hospital. Her Twitter mentions were chaos. Adrenaline pulsed, quickening her heart. Blood flowed to her brain and her muscles. She was alone, but she didn't feel like she was.

She remembers thinking, "What did I do?"

She reread what she had posted. She imagined the minds of the others who read what she posted. What did they hear? She returned to herself, wondered what she could have said differently. She began to presume herself guilty and wondered how she might prove her own innocence.

It took time to reach calm. She had to gather her education, her expertise, her convictions. She had to make sense of the dissonance between what she knew to be true and the mess in her mentions.

Leah didn't leave Twitter. She built a following. She became an ob-gyn, providing care that included Pap smears and routine breast exams, delivering babies at three a.m., and sometimes performing abortions. She spent years online fighting misinformation, being attacked and harassed.

She understands that some people want to see her as a monster. When they believe that, she said, "it's a lot easier to shoot you in the head."

By 2018 Leah was exhausted. She saw a tweet asking her if when she lay down at night, she heard the babies scream. She thought it was ridiculous. Usually, she doesn't respond to posts this ridiculous. But she was tired. She was fed up. She was so angry. She snapped.

"It was one of those moments of like 'I'm just so fucking sick of this. I'm so fucking sick of being attacked. I'm so fucking sick of this ridiculous rhetoric that keeps going around and around. I'm going to use your rhetoric against you. Because you are fucking ridiculous.'"

She tweeted: "You know fetuses can't scream, right? I transect the cord 1st so there's really no opportunity, if they're even far enough along to have a larynx."

The tweet went viral. Leah eventually deleted it, but the cycle had started, and there was nothing she could do to stop it. Conservative news sites covered the exchange, falsely claiming Leah performed illegal abortions by cutting babies' vocal cords. (Her tweet had referred to the umbilical cord.) A week after she posted her tweet, the Utah women's health care clinic where she worked told her not to come in. Leah said the clinic manager had been doxed. Shortly afterward Leah was given a mutual separation agreement and instructed to sign to receive severance or leave without pay. (In 2019 Leah filed a lawsuit against *The Daily Caller* for defamation, invasion of privacy, and intentional infliction of emotional distress. Leah and *The Daily Caller* reached a settlement, and the outlet issued a correction.)

In 2020 Leah moved across the country to serve as medical director of the West Alabama Women's Center, but only a couple of weeks later the state took away the temporary medical license she had been granted. The state's board of medical examiners said that they had denied her license, in part, because of her "public statements related to the practice of medicine which violate the high standards of honesty, diligence, prudence, and ethical integrity demanded from physicians licensed to practice in Alabama." Months later, the commission that reviewed her case found insufficient evidence that Leah had violated any standards. Seven months after she was barred from practicing, she was granted her Alabama medical license.

In the summer of 2022, the Supreme Court reversed *Roe v. Wade,* overturning the constitutional right to an abortion that had been in place for nearly half a century. Afterward many states passed abortion bans, often with very limited exceptions. Alabama's abortion ban, which outlaws the procedure in nearly all circumstances, went into effect. *Guardian* journalist Poppy Noor wrote a profile of Leah that explored how abortion providers had been treated in Alabama before *Roe* fell. When the story ran, Leah expected a storm. She steeled herself.

But the mob didn't come. It's like that sometimes, difficult to predict. Leah received only messages of support. She thought people were being sarcastic. When someone sent something kind, she wondered, "Are you trolling me right now?" People she hadn't heard from in years told her they were so sorry. They told her she was amazing. She almost didn't believe it.

We don't need to experience an explicit threat for the body to activate a threat response. As trauma and violence researcher Sherry Hamby told me, these responses "are getting triggered well before there's actual physical or sexual violence, when you just realize that somebody is your enemy, essentially." On the Internet, the barrier for becoming a threat to somebody has been lowered so much.

Leah didn't need to experience abuse for her body to react. She needed only to fear the abuse. Online, it can often feel like danger is everywhere. Trauma expert Dr. Alisha Moreland-Capuia told me the fear response was designed to be both time- and threat-limited, to help us get out of a dangerous situation and back to a normal baseline. One way to think about trauma, she said, is as "a fear/stress response that doesn't turn off."

It's not just the hate, you see, but the expectation of hate. While we await what we have learned is so often coming. Threats, you imagine, are around every corner.

Bodies

When many people talk about online abuse, they speak about it as though it were separate from bodies. But when we experience these attacks, we react physiologically. Many of us sweat and flush and feel our heart race. Some researchers say emotions are constructed by concepts we use to make meaning out of our body's sensations. Online harassment is an emotional and physical experience. Online harassment can also escalate into other forms of bodily violence.

A study on cyberstalking found that women are nearly twice as likely as men to say "personal physical injury" is their primary fear while interacting online. A global study of online violence against women journalists found that 20 percent of women-identifying survey respondents said that they had been attacked or abused offline in connection with online abuse. As linguist Sally McConnell-Ginet writes in *Words Matter*, words "contribute significantly to non-linguistic actions that may indeed break bones as well as have many other corrosive material effects."

For Leah Torres, the violence is everywhere. It is in her emails and her mentions, it has been outside her workplace and in her community. Always it is in her body. People have filmed Leah outside her work and posted videos online. On her first day at the West Alabama Women's Center, which had previously been burned down and shot at, Leah parked in the clinic's lot, exited her car, and heard people shout, "Hey, doc." She didn't know them. She was dressed in regular clothes, because wearing scrubs is never safe. They knew her because they made it a point to know her. They were waiting.

Since 1977, the National Abortion Federation reports there have been 11 murders, 42 bombings, 200 arsons, 531 assaults, 492 clinic invasions, 375 burglaries, and thousands of other incidents of criminal activity directed at patients, providers, and volunteers. Providers are routinely threatened by phone, text, mail, and online. In 2022 the federation documented a 20 percent increase in death threats and threats of harm during 2021. It found a 229 percent increase in incidents of stalking targeting abortion clinic staff and patients.

Leah said that in her experience, law enforcement has been "basically useless." To demonstrate, she starts with a former stalker.

In her early twenties, when Leah was in medical school in Chicago, a man was obsessed with her. He would call her at all hours.

She would hang up on him, but he would call back. Her landline was always ringing. She reported him to police, but they told her there was nothing they could do.

One night she answered the phone and screamed "What?" into the receiver. He started talking. She offered a few well-placed "Uh-huhs," then put the receiver down on the counter, left her apartment, and went grocery shopping. When she returned, he was still talking.

"Psycho," she muttered to me. "But according to law enforcement, this was perfectly reasonable behavior." Eventually the stalker went away on his own.

Leah said she had a better experience with the FBI, though they were inconsistent. When she was living in Utah, one person online had told her he was going to kill her, and another had sent her a gun emoji. Agents, she said, had determined they were not credible threats. When she moved to Alabama and tried multiple times to contact the FBI in Birmingham about her safety concerns, she said no one returned her calls.

Some women are not comfortable contacting law enforcement. Black women, trans women, and queer women have a long history of abuse by the police. Others don't report online harassment to law enforcement because they believe nothing will be done. Legal scholar Danielle Citron writes in *Hate Crimes in Cyberspace* that law enforcement lacks training and has a history of not taking women's complaints seriously.

When I twice called law enforcement to report my own online harassment, I was not concerned that they would behave violently toward me when they arrived, but each time I was still stunned by how little they appeared to understand about my situation. One officer asked, "Why don't you just write under a pseudonym?"

To be a woman is to be told so often that your conclusions are wrong, that there is a way to solve your problem that you had not considered, that there is really no problem at all.

Shiloh Whitney is a feminist philosopher who, like Amia Srinivasan, theorizes about affective injustice, which she says involves a refusal to cooperate with a person's emotion, to tune in to the world according to that person. Emotions can be powerful social actions, reorienting our sense of what matters in our shared situation. A person's anger or fear or delight, for instance, can redirect others' attention and energy toward the things her emotion is about. But one person's emotion can't accomplish that without cooperation from others: the emotion has to be given "uptake." And uptake is not always distributed justly.

Whitney came to her theory of affective injustice through the example of what she calls "anger gaslighting," in which a person's anger is refused that cooperative response so persistently that she comes to doubt herself about her anger. Not only is her anger blocked from becoming an influential social action, but her own capacity to respond angrily can be injured. Whitney thinks that other emotions can also be gaslit. When a person's fear is persistently dismissed, she may begin to lose trust in herself.

Whitney told me that one time she watched a woman on social media share that she was afraid because a man had followed her home the night before, and she gave a description of the man in the hopes of helping other women in the neighborhood stay safe. In response, Whitney said, members of the group told the woman she was being paranoid.

"Whose anger gets uptake? Whose fear?" Whitney asked. "The answers to these questions tell us a lot about who has power in our world and who doesn't."

If a woman can be convinced to doubt her own judgments, no one has to address the sources of her terror. Even when they are certain of their reality, women learn their experiences of violence are often disbelieved.

When someone tells you that most of the people hurling slurs at you online would never come to your home and physically or sexually assault you, you might respond that some of them would. You might say that this language reflects a culture that has made it permissible for people to physically and sexually assault you. That man casting insults at you online may be doing it for fun, or he may be doing it to impress his followers. He may not be the one to assault you or rape you, but his language suggests he may not care if someone else would or did. His language makes us wonder if a woman in closer proximity to him is subject to all the other forms of bodily violence that his language seems to condone.

Online, we are not all navigating the same forms of physical risk. Some women are there to speak and debate, to share writing and thinking, to educate and self-promote. But many others are online to connect—emotionally, sexually, romantically. These women are engaged in a different project, one that requires different assessments around bodily danger.

Alice Chen is a security analyst who is now married, but when she was actively using online dating apps—platforms proven to be pervasive sites of connection for many Internet users yet notorious for catalyzing online and offline violence—she experienced various levels of threat: coded, brazen, some physically violent. There were the pushy men she matched with. There were the men who fetishized her ethnicity, telling her that they loved that she was Chinese, that they had never "tried a Chinese girl," that they wanted to try her "sideways vagina." There was the man who dropped her off (not all the way home, as that would have been too dangerous) and slapped her on the ass as she exited the car. When she scolded him, he tried again.

"I feel like for men the fear is 'What if I get catfished?'" Chen said. "But I think for women, it's like 'Okay, what if I get murdered?'"

Readers have told me:

"I hope . . . one of them rob you at gunpoint."

"We will not rest until the 'cancerous' scourge that you front for is eradicated for good."

"American citizens know who is under attack. And it sure

isn't assholes like you with your ridiculous foreigner name! You are the racist and the bigot and a huge asshole who is just part of the bigger machine. Be careful you don't go too far."

• • •

Online harassment doesn't need to escalate into a physical attack to wreak havoc on a body.

When a word hits a body, it may not split skin or crack bone, but we can still feel it. Have you ever been shocked into stillness when someone called you a *cunt* online? Did your jaw feel wired shut when they heaved *slut* in your face? Have you ever had a man on the street remark on your body in a way that made you feel hollowed out? What did it feel like inside your skin when words were used as violence against you?

When words are stressful or threatening, they produce physiological responses. You may try to tell yourself they are just words, but your body understands certain words in certain contexts as threats. In the book *Words That Wound,* lawyer Mari J. Matsuda writes that victims of hateful speech "experience physiological symptoms and emotional distress ranging from fear in the gut to rapid pulse rate and difficulty in breathing, nightmares, post-traumatic stress disorder, hypertension, psychosis, and suicide." When your safety is threatened, or when you perceive your safety is threatened, your body has a number of responses that kick in to protect you. Feeling threatened raises your stress level, gets your adrenaline pumping and your cortisol going. It shuts down your appetite, because your body believes it does not have time to eat. It can convince you that you do not need sleep, because you need to stay alert.

The threat could be a bear on the hill, a man at your workplace, or an email in your inbox. The body responds to the shape of a word, the creak of a floorboard, the shatter of glass, the snap of a belt.

Sometimes these threats arrive from unexpected places. When social and cognitive psychology doctoral student Allison Cipriano was in her master's program at Ball State University, she conducted her thesis on rape humor, looking at whether a man would use rape jokes to rebound from a hit to his masculinity. Her institution required her to disclose the purpose of her research at the end of the study, which concerned her, as Cipriano suspected some of her participants might find her research hostile. When her participants saw the purpose of her study, they lashed out, calling her slurs, sending her threats. She told me that during the experience her body was overcome—heart palpitating, hands shaking.

I have also felt my harassment in my body. When the mob came for me, I wrote it all down, because I knew my brain would insist I forget, and I wanted to remember so that one day I could narrate it. I have felt my harassment sitting on the couch between my daughters, our legs braided under sand-colored fleece. I have felt it in the passenger seat of my car, nervously turned toward my husband, wondering if being a man means not dwelling on these things. I have felt it head-down in the grocery store line, magazine photos of obediently posed women an audience to my discomfort. I have felt it in the kitchen with my mother, who, when I read a few aloud, almost choked on lunch before shrieking, "Delete!" I have felt it in rooms with friends, trying to read them aloud, but the kids kept coming in.

When you are experiencing abuse online, even if you try to process your way around it, rationalize your way out of feeling afraid, your body can still go through a threat response. When they launch into their diatribe, when they call you a bitch, when they warn you to watch your back, you tell yourself they do not know you, they cannot find you, but then you wonder if the door is locked, you look to make sure the back seat of your car is empty. When you walk your small dog at night, you wonder if you are safe, if you can speak with other people who are also walking their dogs. You ask yourself,

"Should I be looking around? How far should I go? What should I do?"

While some people suggest words online can simply be ignored, our body's warning systems don't just turn off because the threats we encounter are on our phones. Neuroscientist Stacey Schaefer told me that humans "haven't evolved to adapt to the Internet."

The brain, the human operating system, was not built for the Internet age. Our brains evolve slowly, while technology, in relative terms, has evolved in the blink of an eye. It's still evolving, and our brains have not caught up. This means human instincts are now in overdrive. Things our bodies historically did to protect us from danger "backfire in this new context," said linguist Claire Hardaker, who studies aggression online. Our fight-or-flight response, for example, our automatic physiological response to threats, worked great when a saber-toothed tiger was headed straight for us. That same response system does not work well in an online environment, where some of us are getting multiple abusive comments a day, which our bodies perceive as threats. So now when our hands and feet go cold because our bodies are dumping adrenaline into our blood so that we won't bleed to death from defensive injuries, we are just sitting at a keyboard.

"The job of your brain isn't to keep you happy, it's to keep you alive, and if to do that it needs you to run on the knife edge of anxiety every time you get a message, it's doing its job," Hardaker told me. "It's doing what it was programmed to do. . . . Your brain doesn't have a mechanism to go, 'It's just the Internet. It's fine.' It's working on the premise that there is a threat, in whatever form, and it's coming at you. And that's why the physical reaction is there."

I asked her if our human operating system could ever catch up, if our brains could evolve to adapt to the Internet.

"It would take thousands of years. It would be generations from now. The earth would've spun screaming into the sun," she said.

And even if somehow we could catch up, that assumes technology would stop evolving, which it never will. As I write this book, we are entering the age of AI, which some individuals and groups say, in the wrong hands and absent regulation, threatens to perpetuate gender and racial bias as well as further undermine free expression online. AI for the People CEO Mutale Nkonde, who is pursuing a PhD at Cambridge University investigating how hate propagates online, told me that AI systems are trained using information given to them by developers, and that content can contain knowledge gaps and reflect biases that lead those AI systems to "construct a reality" that bolsters sexist and racist narratives. In a 2023 report, PEN America said generative AI could also make "disinformation and online abuse campaigns easier and cheaper to carry out, at greater volume."

Chronic Stress

I have heard many women wonder what these experiences online are doing to their bodies. I have often wondered what these experiences have done to my own. I have made jokes that my job was killing me, while wondering if my job was killing me. Women online who are constantly encountering threats, anticipating threats, and watching others like them experience threats are cycling through responses that lead to chronic stress. Chronic stress is the problem. Chronic stress can kill you.

Neuroscientist Lisa Feldman Barrett writes in *Seven and a Half Lessons About the Brain* that when someone insults or threatens you, your body budget may be taxed in the moment, but there is no physical damage to your brain or your body. You'll have a physiological response, your heart might race and blood pressure will rise and you'll sweat, but you'll recover. Science shows there may even be a benefit to occasional stress, which can function like exercise. "Brief withdrawals from your body budget followed by deposits create a stronger, better you," she writes. But if you're stressed repeatedly and you don't have the time, space, or resources to recover, then the things contributing to your stress can "gradually eat away at your brain and cause illness in your body. This includes physical abuse, verbal aggression, social rejection, severe neglect, and the countless other creative ways that we social animals torment one another."

Chronic stress lowers immunity and can make us susceptible to some types of cancer. It is linked to inflammation, which can lead to chronic disease. Elevated levels of stress are associated with cognitive impairment. Chronic stress can change the structure of our brains, making us more prone to mood disorders. There is evidence that offspring can be affected by trauma a parent experiences before their birth.

I think about the things my daughters and I experienced together when they were inside me. My former colleague Nicquel Terry Ellis was pregnant and working at CNN when she wrote a story about the media's preoccupation with the cases of missing white women at the expense of those of women of color. It led to the most online abuse she told me she ever received. One person found her cell phone number and texted her "you pple are disgusting animals" and that they hoped she would "choke on a white dick" and wished she would "drop dead." She was eight months pregnant, and she was terrified. She and her baby lived that together, in her body.

Leah Torres said it is obvious that sexism and racism contribute to poor health, but during her medical school training, it was never discussed. Many of her patients at the Alabama women's clinic are Black, and many come from low-income backgrounds.

"If you're walking around stressed out that you're going to be murdered because of the color of your skin, of course you're going to have high blood pressure. Of course your pancreas isn't going to work right. Of course you're going to have these medical problems. Of course you're going to have chronic illness," she said. "In the field of medicine, we have done a disservice to so many people in this country by discounting . . . what it means to live day to day as someone who is not white, as someone who is not a man."

Leah's comments reminded me of a story I reported about microaggressions, whose accumulated impact can affect long-term health. In 1970 Harvard psychiatrist Chester Pierce coined the term *microaggressions* to describe racially charged "subtle blows . . . delivered incessantly." As part of the reporting, I interviewed Arline Geronimus, a professor of health behavior and health education at the University of Michigan, who told me the near constant stress of discrimination and poverty "weathers" marginalized bodies. She reminded me of activist Erica Garner, who fought the system of police brutality that murdered her father for selling cigarettes. Geronimus told me it was possible Garner's body had experienced cellular aging and damage to systems and organs, making her susceptible to health problems and less able to fight them. Garner died of a heart attack at twenty-seven years old.

In 2014 Black feminist activist and writer Shafiqah Hudson helped expose a critical disinformation campaign targeting women of color online with the hashtag #YourSlipIsShowing. While I was conducting research for this book, she died at forty-six years old. Her brother told *The New York Times* she had Crohn's disease and respiratory ailments. Before she died, Hudson told her followers on social media that she had long COVID, had been diagnosed with cancer, and had no money to pay for her care.

Anger

In 2018, when Leah tweeted that fetuses couldn't scream, she was tired. She had been threatened for years, berated for years. She snapped.

Scholar Sara Ahmed calls it the "feminist snap": "When you can't take any more of it, what happens? The moment of not taking it is so often understood as losing it. When a snap is registered as the origin of violence, the one who snaps is deemed violent."

Leah was angry when she snapped. But even when she reflects on what that anger cost her, she thinks her response was fitting. One of the ways she copes with her abuse, she told me, is by giving herself permission to access rage, that historically outlawed emotion that women are taught they are never entitled to.

"I should have been angrier earlier in adulthood, honestly. I feel cheated out of some years of anger," she said. "I use it to give me energy. I use it to give me passion in my work. I use it because my passion is justice, and what better energy sourced for justice than rage."

Many of the women I interviewed for this book told me they were angry. Many of them sounded angry. Our conversations were often measured at the start, but by the end, some women were shouting, others were weeping, many were seething. Sometimes they grew angry when my questions were too simple or missed the mark. Sometimes they were angry at something I asked that was too obvious or at something that suggested that, for all my experience, I still did not understand something essential. Sometimes I was angry at myself for asking questions at all, for resurrecting pain so I could describe it in these pages.

Alexandria Onuoha was angry that her dance program did not value her art, and she was angry that she was abused online for calling on her university to do better. Maria DeCotis told me she is full of rage that men speak to her on the Internet with such disrespect. Online harassment researcher Caroline Sinders, a sexual assault survivor, told me that when someone makes a rape joke online, Sinders is angry that they cannot filter those jokes out, that "in an era of recommendation algorithms and personalization, the Internet is still a deeply impersonal and one-size-fits-none space."

Activist Jaclyn Friedman told me she is enraged about how much labor is required to stay safe online each and every day. "This is my one wild and precious life," Friedman said. "There is so much we could not only accomplish but also experience in terms of connection and pleasure and love if we didn't spend our lives both doing the emotional trauma work and also the logistical work of trying to make ourselves harder targets for some of the worst people out there. It's something I will always be angry about. It's something I don't ever want to take for granted."

Leah has come to view anger as a crucial emotion. She is angry, so she responds to my DM asking to interview her for this book. She wants people to understand the extent of the threats women face. She is angry that in Alabama women are denied access to reproductive health care, so she records a video with the National Women's Law Center to raise awareness about what is happening in her state. She is angry when a pharmacist denies one of her trans patients medication, so she gets on the phone and makes sure the prescription is filled.

As philosopher Amia Srinivasan writes in "The Aptness of Anger," anger involves "a *moral violation:* not just a violation of how one *wishes* things were, but a violation of how things *ought* to be." Anger is an appropriate response to an unjust world. Anger turns us out, has us looking toward sources of violence. Shame turns us in, searching ourselves for culpability in the violence committed against us. Research shows that the more anger a woman feels about her oppression, the more likely she is to take action to address it. As Audre Lorde said in a 1981 speech, "Anger is loaded with information and energy." Do we think millions of women would have marched against the 2016 election of Donald Trump if they were not angry? Do we think millions of women would have tweeted #MeToo if they were not full of rage?

Anger can be animating, but many women have trouble accessing it and have never been taught how to effectively use it. Most of us learn that anger is something to manage and tamp down. Good women do not yell, do not raise their voices, do not snap. Some women are conflicted about how anger feels and whether they want to express it, since many women believe being likable is akin to being safe. And while no woman is really permitted rage, some women are shown they have even less right to claim it. Black women who express anger are perceived as more aggressive and hostile than white women who express the same, and the repercussions of expressing anger may be particularly harsh for poor women. Leah lost a job after she snapped. Not every woman can afford to.

Scholar Stephanie Shields writes that anger "is an emotion of privilege," and "who is socially expected to express anger reflects beliefs about who has a right to the experience." Black women have to make calculations about anger that white women do not. Many Black women suppress their anger to avoid racism and discrimination and sometimes violence. As Srinivasan writes, "Victims of injustice often face a conflict between getting aptly angry at injustice, and bettering (or at least not worsening) their situations. Just what sort of conflict is this?"

Sex therapist Dr. Donna Oriowo, who built a platform that helps empower Black women to explore pleasure and sexuality, talked to me about some of these calculations. She told me that when she begins to move toward anger, she can feel herself weighing its usefulness, becoming "cautious and contemplative." She asks herself, "What do you do with this? What is your next step? Are you going to respond? Are you not going to respond? How are you going to respond? What do you need to be back in your body?"

We are not all afforded the same entitlements around anger. As Soraya Chemaly writes in her book *Rage Becomes Her*, in the United States "anger in white men is often portrayed as justifiable and patriotic, but in black men, as criminality; and in black women, as threat. In the Western world . . . anger in women has been widely associated with 'madness.'" She argues that anger as an emotion is not the problem. Anger is natural, legitimate, and necessary. Problems arise when we do not know what to do with our anger, when it is suppressed or released explosively. Mismanaging anger can lead to health problems, affecting our immune systems and the health of our hearts. Suppressing anger may lead to an increased risk of certain cancers. Chemaly writes that for women to manage their anger healthfully, they need to develop "anger competence," a way to own and harness anger. Put simply, it involves "awareness, talking, listening, and strategizing." She encourages women to develop self-awareness about their anger and to explore what solutions will work for the problem they face. Channeling anger, she writes, could look like a woman saving money to quit a job where she's being abused or harassed, so that she feels safe enough to report. Assertiveness, aggression, and anger all have their place, she says. Women must give themselves permission to stop pleasing people, to be disliked, to rub people the wrong way. Be in community with other women who will validate your emotions, she writes, and never judge another woman's rage.

• • •

Several years ago, before we spoke, Leah gave a talk about abortion at a church in Pasadena, California. During the Q&A, a white woman stood up, took the microphone, and told Leah, "I struggle with rage." Every Friday, she said, a man stood in front of her local Planned Parenthood office with provocative pro-life signs, trying to intimidate women.

Leah empathized with the woman and suggested that perhaps she could shift from feeling anger toward this man to feeling pity for him. Clearly, Leah said, he had filled a void with hate.

I asked Leah, If the woman stood up today and posed that same question, would she answer differently?

"Yes," she said, though she notes that when she made her original statement, she was aware of her audience, the church in which she was speaking, and she was trying to remain sensitive to the room. But times have changed, and in some ways she has changed. *Roe* has fallen. Now, Leah told me, she would tell the woman to "accept your rage, acknowledge it, know that it is justified." She said she still feels pity for the man outside the Planned Parenthood office, but it is also reasonable to feel angry about his intransigence, to feel angry we live in "a sexist, racist society that's built on capitalism and built on owning people, including and especially women." It is not easy to see the world as it truly is.

The anger, she said, is insistent. It appeals, entreats, makes its own demands. It urges: "Keep going. Clearly, we're not done here. There's a lot more that needs to be done."

CHAPTER 4

Thin Skin

THE FIRST TIME I interviewed Minnesota representative Leigh Finke, she was freshly elected and had been on the job for exactly eight weeks. It was February 2023, and Leigh had just become the first openly transgender person to serve in her state's legislature. When we spoke, she did not appear to hold any delusions about how difficult the job would be. She sounded prepared to dedicate herself most resolutely.

Leigh's queerness is visible, and when she decided to run for public office, she had already known punitive violence. She had been assaulted on Minnesota's Green Line train. And during daily commutes to her job at a multimedia publishing company, she had been ruthlessly harassed on the Skyway, a system of second-level pedestrian bridges that connect eighty Minneapolis blocks. Leigh had a presence online, where people intentionally misgendered her. They told her she was a lie. It wasn't a singular denigration. It tracked with a rise in coordinated violence against trans people across the country. Leigh watched as legislature after legislature passed bill after bill denying trans people human rights.

After she was elected, Leigh's online abuse escalated, but she told me she would withstand it. "If I can funnel hatred that is sent at our community toward me, a white person who has institutional power as of January 3, that's part of my job. . . . That's work I can do for my community. I can absorb that. Hopefully, it gives people more space to breathe freely."

In the months after we spoke, Leigh's profile grew, unleashing a new wave of abuse. When she introduced the Take Pride Act, which created new definitions for gender identity and sexual orientation in state law that were more inclusive of queer and trans people, she was targeted by the extremist group Gays Against Groomers, which, the Southern Poverty Law Center writes, "amplifies dehumanizing anti-trans rhetoric" and "perpetuates anti-LGBTQ+ stereotypes by falsely claiming that LGBTQ+ supporters of trans rights are dangerous to society." The group misrepresented Leigh's bill, claiming that her proposed changes created protections for pedophiles. Leigh was sent numerous explicit death threats. Online harassment from her constituents surged.

Seven months after our first interview, and after the legislative session ended, I talked to Leigh again. She spoke differently: "I will not stay in my job if what happened last year continues. I will not survive that."

Off the top of her head, Leigh thought of ten people who live in her district who are committed to harassing her out of office. The abuse, she said, feels claustrophobic.

"People who are from Minnesota and are sending me angry, vitriolic, hateful messages, that does trigger me in a way that's different than anonymous or unknown location people," she said. "This person lives in my city. That's scary. I don't know who they are, but they know who I am. . . . There's just a proximity warning that goes off in my body."

When Leigh read her online abuse, she may have been physically safe, but her body wasn't always sure. Sometimes her body was back on the Skyway. Sometimes it was back on the train. She has not ridden the train since.

Variability

Different lived experiences engender different responses to violence. Our stories of online abuse may share many of the same visible features, but our reactions are distinct, shaped by our identities and personal histories. Our responses are not decided by the proverbial thickness of our skins but are influenced by the deep and rich variability of our lives. I'm tired of aspiring to generate more flesh. It does not bring us any closer to a gentler world. As Roxane Gay wrote in a *New York Times* op-ed: "Who is served by all this thick skin? . . . If we all had the thickest of skins, no one would have to take responsibility for cruelties, big or small."

Intersectionality—a popular framework for thinking about the complexity of oppression and inequality—theorizes that our identities create intersecting forms of discrimination and vulnerability. We cannot easily quantify the impact of multiple marginalizations, because everyone experiences harmful systems differently. On the Internet, our identities often serve as attack vectors: attacked for being a woman, for being a trans woman, for being Black, for being Jewish, for being disabled. Scholar Alice Marwick told me she once interviewed a Muslim person who was inundated with Islamophobic harassment every time they put "Muslim" in their Twitter bio. When you are a marginalized person online, you are targeted not only for what you do but also for who you are.

I will not experience online abuse in the same way as the woman next to me. What moves her may not move me, what terrifies me may not even give her pause. Even when two people share a social identity, their responses will vary, because those responses are shaped by everything else that a person has experienced in their life, every other time they felt as though they were wrong, or not enough, or too much, every kindness they were shown, every lesson they have learned about what it means to overcome difficulty. Many of our reactions reflect our previous experiences of violence and our always-there awareness of the potential for violence.

We are differently vulnerable, but the platforms do not treat us this way. Some researchers argue that platforms should be more sensitive to these differences. In a paper presented at the 2023 Conference on Human Factors in Computing Systems, researchers led by Carol Scott of the University of Michigan called for more trauma-informed social media, arguing that to reduce harm, online platforms must implement "sensitizing concepts" drawn from the six principles of the trauma-informed approach of the Substance Abuse and Mental Health Services Administration, which include safety, transparency, peer support, collaboration, choice, and responsiveness to issues around gender, race, and culture. The authors wrote that harm online is "often overlooked, dismissed, or exacerbated by social media platforms that fail to account for or even acknowledge the systemic disparities that enable them." The research led Scott to partner with UX designer Melissa Eggleston to co-found Trauma-Informed Technology, an organization that advocates for social media companies to design and moderate their platforms with the goal of maximizing psychological safety for all users.

Caroline Sinders is a machine learning researcher who also advocates for social media companies to address online harassment through better platform design, "mitigations" that include being able to remove mentions from social media streams and having consent built into interactions. Sinders is not triggered by rape jokes on the Internet, but they told me that doesn't mean platforms should be free to disregard people who are.

"I have a very high pain tolerance," they said. "That doesn't mean, because I personally have a high pain tolerance for different things, that we shouldn't have Novocain. That's an absurd way to live our lives, so why would we make those same sweeping statements about things like online harassment? Just because I'm not personally hurt, that doesn't mean that's where the barometer of the world sets. What a narcissistic way to view the world."

Retraumatization

Suffering spares no one. Most people experience violence and trauma in their lifetime, but people from marginalized groups tend to experience a disproportionate number of traumatic events, often related to discrimination. "If you are someone who belongs to a marginalized group," says Viktorya Vilk, director for digital safety and free expression at PEN America, "who has had to deal with a lifetime of sexual harassment, potentially sexual assault, racial macro- and microaggressions, potentially assault on the basis of your race or your gender or your sexuality—so you already are carrying trauma—and then you have a string of day-in, day-out messages where people are lobbing hateful speech at you, where they're threatening you, where they're harassing you sexually and otherwise, all that triggers your trauma that you may have built up over a career or a lifetime."

Women experience PTSD at two to three times the rate men do. More than half of women in the United States report some form of sexual violence involving physical contact, and sexual assault carries a high risk of PTSD. People who identify as LGBTQ are at increased risk of traumatization. Women of color experience the chronic stress of racial trauma. Transgender people are more than four times more likely than cisgender people to experience violent victimization.

Leigh has experienced multiple forms of wounding and harm in her private life and public role. Some injuries have been severe, others more enigmatic. She was the victim of a random violent assault for being a trans woman in a public space. After she was elected, the harassment campaign against her was so vicious and engulfing that she called her first year in the Minnesota legislature "traumatizing." There were also the less conspicuous forms of distress. While she is careful to note that her experience cannot be extrapolated to other trans people, she called early transition "dark," a time she was not sure she would survive. She was married, with two children, making a declaration of self that demanded she disrupt every part of her life, managing a nascent femininity, relinquishing the privilege and power of being a man, and the tumbling turmoil and grief left her deeply depressed.

Many women I spoke with told me they found their online abuse triggering and at times retraumatizing. This was especially true for women who engaged in personal, public disclosure—sharing on everything from motherhood to sexual abuse to coming out.

Trans activist Erin Reed said she used to be more open online about the trans experience, but after overwhelming abuse, she doesn't share personal information on social media nearly as much, a shift that has flattened her presence online. Her accounts, she said, no longer reflect who she really is. Now, Reed says, she'll share personal information only if she considers it consequential, a rebuttal or correction made more potent by an intimate perspective. She told me during hearings around anti-trans legislation, when people talked about trans bodies as "mutilated" and "disfigured," she was compelled to share her own experience, writing a Twitter thread on her post-op anniversary, dispelling myths about what these procedures mean for a body. She tweeted, "I am not 'mutilated,' it is not an 'open wound,' and it is fully sensate, despite the best wishes of anti-trans folks."

Reed knew such private disclosures would invite harassment, she told me, and they did, but afterward she also received many messages from other trans women telling her, "I needed to hear this."

I think often about the risks and rewards of private disclosure online. When people bring their full and ostensibly most authentic selves to the Internet, we see more ways of being in the world than our own situations might allow us to imagine. These intimacies can expose the absurdity and limitations of our own experiences. The content many users create through the use of personal disclosure makes new realities possible. It seeds hope. These are extraordinary acts, but they also offer detractors more material to exploit.

While I was researching this book, Nadya Okamoto, a young Asian woman who built her platform by speaking openly about period shame and sexual trauma, showed up on my TikTok "For You" page. I saw a video of Okamoto dancing in her bra and a thong with a squad of #periodpixies, playfully turning her backside to the camera to reveal the wings of her pad. The nearly twenty thousand comments below her video were a mix of praise, confusion, and disgust.

I couldn't help but return to high school, when I seeped through my pants, when I had no sweater to tie around my waist, when I walked alone from classroom to bathroom, feeling the stickiness of shame, because the most important thing most girls learn about their periods is how to hide them. Many years later I would hear then presidential candidate Donald Trump malign journalist Megyn Kelly, saying that she had "blood coming out of her eyes, blood coming out of her wherever."

Watching Okamoto, I saw a woman refusing a narrative of shame, doing something with her body that I thought I never could, absorbing messages of disgust so that perhaps a young girl might decide her period leak was not the end of the world. I thought about how much accounts like this matter to people, and I considered what possibilities we lose when those who run them are bullied into silence.

There is an entire "snark" subreddit dedicated to picking apart Okamoto's life. Okamoto told me she can generally brush off commentors maddened by her refusal to hide her menstruation, but when people attack her over her disclosures of sexual abuse and assault, it's much harder to set aside. Okamoto is a rape survivor, and she has also spoken about sexual abuse that she says she experienced in childhood. I have watched videos of her where she says things publicly many survivors struggle to admit to themselves.

When I interviewed her, I asked, What are the most difficult experiences for you online?

The ones that resurrect the pain of not being believed, she told me.

"The nature of talking about my trauma online is that people have a lot of feedback," she said. And while it is the cost she is willing to pay to fight society's injunction to silence, it is still difficult to dismiss other people's doubt.

"It's triggering when I get online and people are like, 'Oh, but that didn't happen.'"

When your disclosures are met with skepticism and denial, when people tell you your suffering is a fantasy, it can return you to the tenderest parts of victimhood.

In 2007 activist Jaclyn Friedman, a rape survivor, wrote a piece for the nonprofit online news service Women's eNews on how to address the connection between drinking and rape without victim blaming. She also offered concrete social and institutional interventions that could help prevent sexual violence. She included a paragraph about her own experience, which went viral in the way stories could back then, linked to on various feminist blogs. Friedman said she was shocked and unprepared for the cruelty. Many people online questioned whether she was actually a survivor. It was infuriating that people could dispute a truth that her body knew absolutely, and that there was nothing she could do to stop it, that online those lies would form a permanent record of their own.

"Fundamentally violence is about powerlessness, imposing powerlessness on somebody else. It did feel like a profound violation. It was incredibly triggering. I was in tears," she said. "I was nonfunctional for days."

A decade after her victim blaming essay went viral, Friedman again found herself at the center of a large-scale attack, this time in the age of social media. It was 2016, Trump had just been elected, and after it was announced that Steve Bannon would serve as a senior adviser, Friedman posted several tweets imploring people not to normalize the decision and referred to Bannon as a Nazi.

Friedman, who is Jewish, said she was then deluged by self-professed Nazis for a solid forty-eight hours. They told her: "We have our eyes on you."

We were talking about this when Friedman told me she needed to breathe. I told her all right, let's do that. We were on the phone, and I couldn't hear her breaths, just silence. She told me she was lightheaded and that it was difficult to speak. I told her we didn't have to. She was quiet for a long time, but I didn't want to hang up. Eventually, she came back to the line. She finished the interview.

Histories

It is not only the daily indignities of discrimination and experiences of trauma that can shape our responses to violence online but also those more quotidian experiences that construct us. It's possible our reactions to online abuse are also informed by moments embedded deep in our histories.

Critical social justice scholar Loretta Pyles told me "underlying our triggers can also be a collection of experiences and imprints from our very earliest childhood memories. Whatever's happening in an online space that's abusive is hitting on old stuff, too."

I don't want to unnecessarily dramatize my old stuff. But I can tell you what might be plausible. I share these memories not because they are direct lines to this present, but because they are possibilities, and maybe hearing them will allow you to make space for your own. I don't know if many of the messages I fixated on would have captured my attention had they not reminded me of some previous discomfort or shame.

In 2020, after an appearance on CNN, one viewer suggested I deserved to die of COVID for my analysis of what it means to be an American: "You piece of shit. The comments you made regarding

America & 'white people,' basically just labeled us all as inherently racist. You fucking bitch! You race-baiting liberal trash. If you were to perish from Covid, it would make the world a better place."

That didn't upset me. What upset me was a separate comment that suggested my lighting was unflattering.

I can tell you that when I was a girl, I was called ugly enough times that it began to count. One of my firmest memories is of being in school, making what I thought was an innocuous comment, and a boy suggesting I was too hideous to have such an opinion. After the CNN interview, I bounded up the stairs to find my husband and asked, "Did I look bad on TV?" I wonder if the child in me just wanted to know: Am I too disgusting to speak?

When I wrote about sexism and people told me to shave my legs, my armpits, that I was too manly to wear a bathing suit, maybe I would have easily deleted those messages, except I remembered this unoriginal refrain, the many times other children suggested I had not sufficiently tamed the fuzzy upper thigh that my mom refused to let me shave or the unibrow I had not thought to divide. One time on the school lunch line, I told a boy not to cut in, and in front of the crowd he spat, "Why don't you shave your fucking mustache." I went home and introduced a Gillette razor to my face.

For many of us, hate comments demand attention by revealing the banality of timeless insults. Sometimes they pick at old wounds. We recoil at their stubborn familiarity, their broken-record quality, the ways they share contours with the violence of our past. Violent language is with us all our lives, morphing to fit our ages, the season, our particular indiscretion. These words come leaden.

Logging Off

When the online abuse overwhelmed Leigh Finke, she minimized her exposure. In some digital spaces, she eliminated it altogether. She did what so many of us are told to do: she logged off.

When Leigh became a politician, occasional Twitter comments became piles of emails. Online abuse that was fairly impersonal became extremely personal. What was once frustrating grew terrifying. When she was elected, she gained a staff, and she used them to create barriers between herself and the abusive content. She stopped reading all comments, and her emails were vetted. She would write her own tweets, but someone else would post them.

"I don't interface with the unknown public almost at all," she said. "I don't answer my phone. Someone pretty much screens my emails. I am subject to such a voluminous amount of online vitriol. . . . It's just too much at this point, at least for me. . . . I don't want to look at it. I don't want to know about it. It's too emotionally compromising for me to even dip my toes into it."

According to the American Psychological Association, avoidance coping is "any strategy for managing a stressful situation in which a person does not address the problem directly but instead disengages from the situation and averts attention from it."

Psychologist Seth Gillihan has said that in psychology there tends to be a bias against avoidance coping, and in some cases for good reason. Avoiding things that are not actually dangerous tends to increase anxiety in the long term and prevents a person from learning that they can cope with the things they're avoiding. It can also lead to loneliness, isolation, and unhealthy forms of escapism. These downsides, Gillihan notes, don't extend to things that are unequivocally harmful—dangerous dogs, dark alleyways—or deeply stressful, with no real redemptive value. Avoiding bungee jumping, he said, would not be considered maladaptive, nor would avoiding impossibly difficult people.

The bias against avoidance coping, Gillihan told me, is a problem when it fails to take into account how intense and merciless certain stressors can be. It comes down to trade-offs: deciding what is best for right now and what is best for the future, what is worth the emotional risk and what never is. Lives change, risk profiles change.

Maybe we need to avoid something today that we might not need to avoid tomorrow. Avoiding exposure to something can work well for specific goals. "It's not that there's ultimate and overriding value in always approaching difficult things," he said. "It's really about whether it's worth what it costs." Leigh needed to minimize her exposure to abusive messages online so she could do the legislative work that had a bigger and broader impact on the trans community.

Influencer Nadya Okamoto told me she has steadily reduced her exposure to toxic content online. As her platform has grown, so has her ability to delegate screening: "I am in more of a fortunate position where I can hire an assistant to manage comments for me. . . . That's been really important to have some separation."

While some women have formalized vetting structures, others arrange an informal system of screeners so they can log off during overwhelming harassing situations. When activist Jaclyn Friedman was inundated after her Bannon tweet, she said she did what she always does: "I asked beloveds to monitor it for me."

Disengagement can minimize exposure to harmful or triggering content, but it can also have drawbacks that are epistemological.

In Alfred Archer and Georgina Mills's paper on emotion regulation and affective injustice, they examine the idea of "attentional deployment," which involves arranging one's attention to influence one's emotions. It can mean looking away from something that is causing you to experience an emotion you don't want to feel. Archer and Mills argue that while this may help limit painful experiences, it also puts women in "a worse epistemic position" by reducing their "awareness of the nature of sexism and how it operates." Turning away from sexist behavior means women are less tuned in to "the various forms sexist behavior can take." They may also deny themselves information that could be helpful in managing basic safety (who are the people to be most wary of?) and reduce their access to resources that might help them cope or even challenge their perpetrators. Archer and Mills are careful not to suggest victims are obligated to expose themselves for the sake of knowledge, but they stress there are many costs to turning your attention away from forms of injustice.

Disconnection may be the most obvious cost of logging off. Many of us go online to meet like-minded others, to foster and nurture friendships, to challenge the limits of our physical communities and proximate social spheres. The pandemic underscored how valuable the Internet can be for connection and showed many of us how significant social platforms have always been for certain communities: trans people, queer youth, those living with disabilities. The Internet can be a locus of lifesaving validation, a home for our most creative and unruly selves, a place to abandon insecurities and sometimes even our vanities.

Leigh said that logging off initially worked well for reducing her exposure to hate, but her withdrawal from social media meant she also lost the ability to communicate with other transgender people: "I loved Twitter. The trans community on Twitter was, for me, just a thriving and wonderful space."

Leigh missed that, and practically, she also had a difficult time staying up-to-date with anti-trans attacks. She called the withdrawal from all social media an "overcorrection."

"I cut off my daily flow of information from other trans people who are doing the fight in their part of the country. Not having that access, that took something away from me. It made me feel more cut off and more isolated because this is lifelong movement work. Knowing that other people are running the marathon with you is really helpful."

Leigh decided to lean back in, albeit differently. Now she is in a Signal group with other LGBTQ, gender-expansive and trans lawmakers, where they empathize and strategize together.

Leigh told me she often wonders, "What is a permissible level of violence?"

One afternoon, Leigh was walking out of the House chamber when an older man approached her and said, "You will go to hell.

You are a demon. You work for Satan. You turn boys into girls. You mutilate their bodies."

It was like online trolling in real life, she said. She could barely compute it. How do you log off from that? "That is an absurd human experience," she told me.

Leigh said she will do the work as long as she can. One afternoon she wrote: "Every trans person deserves a future."

A staffer tweeted it.

CHAPTER 5

Harm

MY SECOND DAUGHTER was born in the fall, in a season marked by paradox. Autumn is spectacular in its decline. I walked my baby into our home under a pure blue sky as deep-veined leaves crunched beneath me. The trees were not naked yet but would be soon. For months I bundled her against a clarifying cold, her body fat with sleep.

Winter took over, and I had just returned from maternity leave when news broke that stunned a nation. On January 26, 2020, basketball great Kobe Bryant was traveling with his thirteen-year-old daughter, Gianna, when their helicopter crashed near Calabasas, California, killing them and the seven other people on board. Bryant was a legend, and it was said that Gianna had the makings of a prodigy. But talent seems beside the point. A father and his child had died on a hillside.

It was a staggering loss that Bryant's family, fans, former teammates, and competitors mourned deeply—but the event also challenged our ability to hold more than one thing as true. Bryant was brilliant on the court, ostensibly a devoted father to four girls, and a fierce champion of women's sports. But nearly seventeen years before his death, he had been charged with raping a nineteen-year-old woman. During pretrial hearings, his defense lawyer repeatedly attacked the character and credibility of his accuser, who received death threats. Prosecutors dropped the case after the woman refused to testify.

Bryant later issued a public apology, which his lawyers said the woman insisted on "as a price of freedom." It was reported that the wording of the statement was negotiated. It read, in part: "Although I truly believe this encounter between us was consensual, I recognize now that she did not and does not view this incident the same way I did. After months of reviewing discovery, listening to her attorney, and even her testimony in person, I now understand how she feels that she did not consent to this encounter."

When women online tried to address Bryant's complicated legacy, attempts were made to silence them. Many were told it was "too soon." But for those who survive sexual violence, accountability is rarely "too soon." It is often "too little." It is always "too late."

Hours after Bryant's death, former *Washington Post* reporter Felicia Sonmez tweeted out another outlet's news story detailing the rape allegations against him. Afterward she was mobbed online and had to go into hiding. *The Post* suspended Sonmez for her tweets but days later reinstated her, with newsroom managers admitting they had been out of line.

For all the conversations social media has made possible, Bryant's death seemed to underscore its limitations. It is not easy within time and character limits, in a world of provocative clips and bite-sized takes, to have robust, nuanced conversations about harm, least of all in the midst of a tragedy. Bryant's death revealed differences in our cultural beliefs around how death should be observed and what it means for damage to be repaired. White women who tried to speak were reminded of their long and brutal history of complicity in perpetrating harm against Black men, especially through false accusations of sexual violence. It is true that no white woman could possibly appreciate what Bryant meant to Black athletes, Black children, Black families. But even Black women who tried to address Bryant's legacy faced harassment.

When *CBS This Morning* anchor Gayle King brought up the rape allegation in an interview with WNBA player Lisa Leslie, King was attacked online, received death threats, and had to travel with security.

As scholar Amira Rose Davis wrote in *The New Republic,* "It's incoherent, putting pieces of a puzzle together that don't fit, jamming them, or discarding one or the other. But we need to hold them. Even if they are heavy. This is the work."

It can be difficult for all of us to collide together in digital space, to bring with us our identities and experiences, cultures and values. Those who believed in Bryant's redemption clashed with those who saw him as a symbol of how money and power can protect a person from consequence. As culture writer Evette Dionne said in a piece for *Time,* "Thanks to the pressures of social media, in which we react to unfathomable news in real time, we often fall into a binary of good or bad, wrong or right, on the side of survivors or on the side of a rapist. It is rarely that simple."

The day after Bryant died, I wrote a story on the abuse women journalists face. I wrote about Sonmez and my *USA Today* colleague Nancy Armour, who was also vilified online after writing in her column that a full account of Bryant's life was necessary and that "to ignore that, or shout down those who won't, puts Bryant in a neat little box where he doesn't fit."

My story did not address how race impacted the charged moments after Bryant's death. It should have. When it published, readers sent me dehumanizing messages. They didn't just tell me I was insensitive or callous or even a cunt. They told me I was nothing. I was sitting in a rocking chair with my baby when I read the message that said: "I am sorry to those who have confused you to be a person, because you are not a person."

Journalists are powerful. The stories we write shape understanding, shift culture, influence elections, expose injustice, and sometimes make certain forms of justice harder to achieve. Social media has allowed readers to fight narratives they deem harmful and dehumanizing. It's possible this user saw me as misusing my own power. But what he could not see through his outrage was my own humanity, the same transgression he accused me of in my coverage of Bryant.

I reached out to my colleague Nancy Armour, a much more visible journalist whose column had provoked an intense online pile-on, to tell her I was sorry for what I knew she was going through, to make sure she was okay. I emailed her.

She responded quickly, curtly, waving me off. "Eh, I've gotten used to it," she said.

I sat there awhile. This response did not align with my understanding of online abuse. It did not align with my experience. Was this resilience? Accommodation? I had gotten a few dozen degrading messages and was anguished and afraid. Nancy was at the center of a large-scale attack, yet she was at the Super Bowl in Miami, seemingly impervious.

Four years after our exchange, I reached out to Nancy again, to interview her for this book. She told me she is rarely *hurt* by online abuse. She told me she isn't bothered by the nasty things that are said to her. When people eviscerate her over her columns, when they tell her to get back in the kitchen, when they ask her to tell them what her husband thinks, when they say, See, this is why women shouldn't write about sports, she isn't crying into her coffee, looking over her shoulder, questioning her work. She is rolling her eyes, shaking her head. She is writing.

When Nancy was mobbed over her Bryant column, she sat in the Super Bowl press room, read some of the messages, and moved on. Not because she was unfeeling, she said, but because she believed in what she had written. She told me she thought the public "collective amnesia" about Bryant was "ridiculous." The emails kept coming, the Twitter notifications, too, and while she could not completely ignore them, she went on with her work. She mingled with colleagues she enjoyed spending time with, caught up with folks she hadn't seen in a while, went for a run with a friend, and covered the circus that is the world's most-watched sporting event. She did her job. She slept well.

Nancy has written on the intersection of sports and culture for *USA Today* since 2014. She told me there is no way to prepare for the experience of abuse online, which for her began when she started writing her column, but she was not caught off guard. Women who cover sports have always faced scorn and ridicule. Nancy does not believe her abuse is acceptable or inevitable, but she takes online attacks with "a grain of salt." On a day-to-day basis, she told me, she is seldom emotionally affected.

Many of us know women like Nancy. Maybe we are like Nancy, or wish we were. But I don't think the lesson here is that some women are fragile and others thick-skinned. I believe the lesson here is that there is a difference between the way we think about the concepts of *hurt* and *harm*. Online abuse doesn't need to consciously hurt or offend a person to cause individual or collective damage.

We need to tell Nancy's story, not because she is a model for how to deal with abuse online but because the absence of hurt feelings makes space for a deeper conversation, a more urgent one. Some people want to focus attention on women's responses to violence, so they may shift attention away from the sources of violence, from the myriad ways it harms the individual and social body. The idea that women should ignore or endure abuses online suggests these abuses only emotionally hurt. That if we could somehow toughen up and feel or express less emotional pain, then we would be all right. If only we could be less sensitive, then we would stop complaining so much.

As philosopher Lynne Tirrell told me, "Learning not to care about the damage isn't the same as not being damaged."

Individual Harm

There is no consensus on the difference between hurt and harm. The psychologist, the philosopher, the sociologist, the legal scholar, the neuroscientist, the average woman will all offer different perspectives on the distinctions between the two and whether there is a distinction at all.

In these pages, we will think of hurt as the conscious, subjective experience of emotional injury, and of harm as damage, an impairment that is longer lasting, sometimes permanent.

Enough hurt over enough time can create harm, especially for the most vulnerable. Surveys show that purposefully misgendering trans or nonbinary people is hurtful, and research shows it can cause longer-term damage, leading to depression, anxiety, and trauma. But even without hurt feelings, damage can be done.

As neuroscientist Lisa Feldman Barrett explains in *Seven and a Half Lessons About the Brain,* chronic stress can produce physiological changes that weather people over time, and our bodies can become sick without ever consciously experiencing emotional hurt.

"If you're exposed to verbal nastiness continually for months and months or if you live in an environment that persistently and relentlessly taxes your body budget, words can indeed physically injure your brain," Barrett writes. "Not because you're weak or a so-called snowflake, but because you're a human. Your nervous system is bound up with the behavior of other humans, for better or for worse. You can argue what the data means or if it's important, but it is what it is."

Amnesty International has documented women around the world reporting stress, anxiety, and panic attacks as a result of their experiences with online abuse. It did not say women were sad; it showed they were harmed. A 2021 literature review by researcher Francesca Stevens found that victims of cyberstalking, harassment, and bullying experienced depression, paranoia, stomachaches, PTSD, and heart palpitations. She found as many as nine studies that showed victims engaged in self-harm or had suicidal thoughts. Some attempted suicide.

Harm can be psychological, physical, reputational, economic.

Harm can be losing a job. After Dr. Leah Torres was castigated online for responding to what she saw as a bad-faith tweet about abortion, she lost her position at a Utah women's health care clinic, a job she loved. "That really tore at my soul," she said. These disruptions and derailments to women's work have been referred to by scholar Emma A. Jane as "economic vandalism."

Harm can be experienced when we choose exposure to violence over the loss of career connections. Comedian Maria DeCotis would like to avoid the most disgusting and abusive things in her DMs, but she feels she has to read anything that comes through on social media, because it could be a professional opportunity. Men in the industry, powerful men, reach out to her on these platforms. "When people are like 'Oh, just ignore them,'" she told me, "it's like 'Oh, I thought this was a professional connection, and now this person is sending me dick pics.' I rely on this platform for my career. I can't ignore everything. I can't just block everyone."

Harm makes it harder for us to function in the world, to fulfill our potential. Some of the individual harms of online abuse are easy to see—the lost job opportunity, the reputational damage, the increase in anxiety medication. Other harms are harder to notice and define—the slow erosion of our bodies through chronic stress, the changes to our self-concept, the heightened cynicism about the wider world. We don't always or immediately notice the ways in which we have become heavier, less hopeful. Quieter.

We can also be harmed by things we are not conscious of, which is another reason harm goes beyond hurt feelings. A muted user who tweets misogynist things at you that you don't ever see doesn't hurt you, but you're harmed because these words shape the attitudes of a culture, and those attitudes shape how you are treated—on the Internet, in your home, by your spouse, by your government. Your blocks and mutes may save you emotional hurt, but other women see those words and can be hurt and harmed by them. You know those words are still there, the language that forms a world in which we don't exist without violation, in which we struggle for dignity, for authority, for the fulfillment of basic needs.

"All the fights about whether something counts as harassment or abuse or whether other people should care are sidestepping the fundamental thing," Columbia University research scholar Susan McGregor told me. "There's harm. Are we going to address the harm or not?"

It is impossible to measure just how much harm has been done. Even when women are conscious of being harmed, they may not want to admit it. Sometimes it makes us feel weak, and sometimes we are justifiably afraid these admissions will make us more attractive targets. There are endless ways to minimize these experiences. We must also make space for the possibility that even women who say they are unaffected, who say they are escaping these experiences relatively unscathed, may be imprecise in their evaluations.

When we narrate our own experiences, we may not ask all the questions an objective party would.

As psychotherapist Abra Poindexter told me: "You might just be like 'Well, I'm used to it,' but that doesn't mean that you aren't drinking more, that you aren't needing more, that you aren't needing less, that you haven't changed your eating habits, that you aren't overexercising, that you aren't having difficulty in trusting men in relationships, if you happen to be in one with a man. You might be distracted, have difficulty focusing, feel more down or more anxious, you might notice simply feeling unsafe and not know exactly why. What's the rest of the story?"

Mainstream conversations about online abuse tend to focus on the most egregious examples, which can eclipse the impacts of subtler forms of harm. Philosopher Lynne Tirrell's early work focused on hate speech, but she came to desire a broader category that made space for an understanding of language that harms gradually, sometimes unknowingly. She developed the concept of "toxic speech," which she writes "comes in many varieties, can be chronic or acute, can damage individuals or society, in whole or in part, permanently or for a time."

A woman does not need to experience the most violent forms of language to experience harm. The words need not be assaultive, just based on presuppositions that challenge our capacity to function in the world. The language doesn't have to feel like a punch in the nose, Tirrell writes. Pervasive toxic misogyny can be experienced "without awareness of the cumulative effect."

Tirrell writes that the harm caused by toxic speech can vary based on the susceptibility of the subject (whether we've been weakened by this kind of speech before), type of exposure (one-on-one speech versus public speech, written versus oral, whether the speaker is a stranger or someone we know), and dosage (how frequently we experience it, and how potent the word or phrase).

One of the reasons I find her metaphor such a useful way to think about harmful speech is that it allows for changes in a person's capacity to cope. Tirrell writes, "We must attend to speech that harms gradually, by changing the boundaries of the normal, and by corroding our capacities to function at our highest levels and to our best-held ideals."

Writer Lyz Lenz said that one of the comments that impacted her most was neither profane nor explicitly violent. It was 2019, and Lenz was at work at *The Gazette* in Cedar Rapids, Iowa, when she logged onto Facebook to see what was happening in her community. She read a comment from a local woman. She told me it went something like this:

"I saw Lyz Lenz out at dinner the other night. She was sitting at the bar with some friends talking and laughing really loud, wearing an inappropriate outfit and swearing. And I just don't know how she can be considered a representative of the newspaper if her behavior is so awful."

Lenz thinks she was wearing skinny jeans and a V-neck.

"Of all the things that have happened to me, that Facebook comment probably had the biggest impact. It was just a dumb little chiding, which is so easy to make fun of and move on from. But it just hit me at that moment where I was like, 'Nowhere is safe.' Because a lot of times people tell you that it's just an online problem. . . . She took my ability to be a normal person in my town away from me. And maybe she didn't take it, but she was the one who told me it was gone."

Societal Harm

Online violence doesn't just harm the individual woman. It is a violence that shudders through society. Harm can be a narrowing of the ideas we are willing to share and the stories we are willing to tell. To avoid abuse, we censor ourselves, limit ourselves, sometimes leave online spaces altogether. Research has found that 41 percent of women ages 15 to 29 censor themselves online to avoid harassment and that the self-censoring of female scholars, journalists, and politicians creates a "digital Spiral of Silence." Nearly two out of three women journalists said they'd been harassed online at least once, and of those, approximately 40 percent said that they avoided reporting certain stories as a result. As former *Washington Post* columnist Margaret Sullivan wrote in 2021, "Unless you've been there, it's hard to comprehend how deeply destabilizing it is, how it can make you think twice about your next story, or even whether being a journalist remains worth it."

Harm can curtail the questions we are willing to ask. Social psychologist Allison Cipriano was abused by her own research participants for a study on rape jokes and masculinity. What research isn't being done, what questions about the world are we not answering, because the people who want to undertake that research are afraid for their safety?

Dismissing online abuse as nondescript nuisance or individual horror obscures its impact on culture and politics. It trivializes the organized, tactical nature of many abuse campaigns. Sexual violence researcher Nicole Bedera said that there's a disconnect when she tries to talk to white men about her experiences of harassment. They look at the words on the screen and think, "That's no big deal." She had to explain: "The violence isn't necessarily just the comment, the harassment in the moment. It's the structural violence across our society."

Online abuse of women is not a personal problem, even when it looks and feels that way. PEN America's Viktorya Vilk said that what is so ingenious about online abuse as a tactic to intimidate, censor, and silence is that it is made to look and feel personal. It seems like it's about you.

"It's then very easy to not treat it like a systemic problem that requires systemic solutions. If we don't zoom out and understand that there are antidemocratic forces at play that are trying to push women out of public office, out of public life, out of the journalism industry and many other industries, and we don't tackle that at scale, the tactic will have worked, which is that we'll just have way less women and also way fewer people of color and LGBTQ folks on the Internet and in public life."

Violence online is a social problem, the problem of a white supremacist, patriarchal society, a political problem, a movement to set back the rights of women and people of color and LGBTQ people.

When I was researching this book, at some points when I was trying to justify it even to myself, I kept returning to the 2020 speech Representative Alexandria Ocasio-Cortez made after Representative Ted Yoho allegedly called her a "fucking bitch" on the steps of the U.S. Capitol. Afterward she said on the House floor:

"In front of reporters, Representative Yoho called me, and I quote, 'a fucking bitch.' These are the words that Representative Yoho levied against a congresswoman. A congresswoman that not only represents New York's 14th congressional district, but every congresswoman and every woman in this country, because all of us have had to deal with this in some form, some way, some shape, at some point in our lives. And I want to be clear that Representative Yoho's comments were not deeply hurtful or piercing to me. . . . I have encountered words uttered by Mr. Yoho and men uttering the same words as Mr. Yoho while I was being harassed in restaurants. I have tossed men out of bars that have used language like Mr. Yoho's, and I have encountered this type of harassment riding the subway in New York City. This is not new and that is the problem. Mr. Yoho was not alone. He was walking shoulder to shoulder with Representative Roger Williams. And that's when we start to see that this issue is not about one incident, it is cultural. It is a culture . . . accepting of violence and violent language against women, and an entire structure of power that supports that."

Ocasio-Cortez made a point of spotlighting Yoho's words not because of individual shock or hurt but because she recognized the words' power to collectively harm. Her speech stressed that this language is omnipresent, that it is reprehensibly normalized, that it is said by men with power, men who have wives, men who have daughters, men who claim to be "decent." Ocasio-Cortez said Yoho did not just call her a "fucking bitch." He called her "disgusting," he called her "crazy," he called her "dangerous."

She did not make a speech to address a personal insult. She made a speech to address a culture of white male supremacy in which women are routinely denigrated, humiliated, and silenced. She would not be silenced.

Consciousness-Raising and Counterspeech

When Nancy Armour says she isn't hurt by experiences of online abuse, we can only speculate why. We can imagine that her reactions and responses are informed by her identity, her history, her personality, her years of experience, the institution behind her, the other women sportswriters in her corner. It is also possible that, like many women, she is habituated to some of her abuse.

We survive because we adapt. It would be disingenuous for me to suggest I have not also grown accustomed to some things. I don't talk about my abuse nearly as much as I used to. The individual instances don't stun or pain me quite the way they once did. I intellectualize around them. But the accumulation of them, the expectation of them, still requires work I would rather be engaged in elsewhere.

I wanted to interview Nancy not only because I was curious about her reactions, their departures from my own, but also because I was interested in the way she visibly responded to her abuse. When I emailed her over the Bryant column, it was after I saw her retweet a message from someone who called her a "cunt."

Nancy, like many women online, has a history of publicly sharing her abuse through retweets and screenshots, often with commentary. She doesn't always file the harassment away, delete it, or let it quietly pollute her mentions. She has publicly referred to her abuse as "fan mail," mocked her abusers' insults and their punctuation, questioned their lack of creativity. Sometimes, she says, she wants to present "evidence" that shows people, "here is what we are seeing on a daily basis," which she hopes might make it harder to suggest a woman is exaggerating, harder to accuse her of "being too sensitive."

Psychotherapist Seth Gillihan has said that exposing online abuse, like exposing other forms of violence, may be a way for a woman to reclaim control by putting a frame around what is happening.

"Random disturbing messages can feel more dangerous and less controllable than ones that have been captured, described, shared, and maybe in a way contextualized," he told me. "The woman is not a helpless victim receiving these messages but can repurpose the messages to refute them, expose them, have their reactions validated by others perhaps. There can be a lot of value in just marking unacceptable behavior as unacceptable."

While the act of calling out abuse may feel personally empowering for some women, these broadcasts are often also focused on exposing how oppression functions in vicious, vivid detail. Nancy says it's difficult to leave hate unaddressed. She worries if she does not condemn abuse, especially attacks her readers see, it may send the message that she accepts mistreatment. Screenshotting abusive messages or even retweeting them is a way she calls out the energy behind *cunt,* behind the bodily threat, behind the dressing her down.

"It is important to call it out because if you don't, then it gives the impression that it's okay, and it's not okay," she said. When she reflects on some of the abuse she's received in response to her supportive coverage of trans athletes, she considers: "If an LGBTQ person sees that and there's no response, does that cause them more harm?"

Scholar Mindi Foster's research has found that publicly tweeting about societal sexism often includes the goal of mobilizing support for change. Many women I interviewed who felt compelled to expose discrimination and abuse hoped it would do something—change a policy, at a minimum change a mind. Alexandria Onuoha tried to expose the violence of white supremacy on her college campus through her op-ed, which she hoped would result in concrete changes to her dance program. Comedian Maria DeCotis exposed magakyle's dick pics both as an effort to manage her own safety and to educate people on the harms women comedians face. Women are stressing they do not want to live this way.

Nancy told me she often shares abusive messages with the intent of moving people in positions of power. "When I do call people out or when I do retweet something that I've gotten, that's my attempt at . . . making people inside the business and outside the business think," she said.

In 2016 Chicago-area sportswriters Julie DiCaro and Sarah Spain made a PSA about women's online abuse that exposed some of their most venomous harassment. The reporters had men they did not know read the harassment to their faces. The men are visibly uncomfortable; they sigh and pause and shift in their seats. Sometimes, after the men read some of the worst words, they look at the women expectantly, searchingly, as if to ask, How can you live like this? Other men say they are having trouble making eye contact.

They scan the room, looking, it seems, for answers or maybe for someone to stop it. They say, "I'm sorry."

DiCaro and Spain created the PSA to encourage people to make different choices about their behavior online. It has been viewed nearly 5 million times.

There was no consensus among women I interviewed on whether the strategy of exposing abuse was useful or self-sabotaging. Baring individual instances of hate risks amplification and could arguably create vicarious trauma for other women on social media platforms who are exhausted by the toxicity. Amplifying abusive messages, whether by retweeting with commentary or directly responding, may also give many abusers the attention they desire. Calling out bigotry also comes with risks not every woman can afford. Some of the risks are personal—I always consider how my speech may affect the safety of my young children, and at least one woman told me that now that her children are grown, she finally feels comfortable publicly challenging her antagonists. Other risks may be professional, like alienating colleagues or potentially losing a job. Some news organizations prohibit reporters from speaking publicly about their harassment.

One also has to wonder whether those who have the power to do something about online abuse—employers, social media platforms, lawmakers—need any more evidence of the violence we face.

There is an efficacy question here that demands special attention, as in practice iterations of this strategy fall under the umbrella of counterspeech, which U.S. Supreme Court doctrine has historically favored over censorship, except in cases of incitement. Counterspeech seeks to undermine hate with a different narrative, and proponents of the strategy argue it can mitigate community harms as well as improve public discourse. Skeptics of counterspeech wonder whether fully free expression and equality can coexist.

The Dangerous Speech Project, which studies speech that inspires violence between groups of people, defines counterspeech as "any direct response to hateful or harmful speech that seeks to undermine it." Cathy Buerger, the organization's director of research, told me that by "direct," the organization generally means a response to a specific piece of speech or text, which is sometimes seen by the person who created that speech but not always. Counterspeakers may respond to a wide variety of forms of harmful expression, including hate speech, dangerous speech—defined as any form of speech that condones or leads to violence—or dis- and misinformation.

There are different theories on what qualifies as counterspeech. Philosopher Lynne Tirrell told me her conception requires a response that counters an assertion or characterization, which can come in the form of words but can also include images or memes. Reposting an abusive message without commentary wouldn't meet her definition. "Without commentary the repetition of the tweet . . . is pure amplification," she said, noting that counterspeech should include content that "directly counters" what the original speaker said. If David calls Maria a cunt, for example, it's not enough to say "No she isn't!" which only counters the application of *cunt* to Maria, leaving the term uncontested. That kind of limited response keeps the nasty language alive and only changes who it can be applied to. Calling women cunts is a common practice that also needs to be countered, she argues, so in her conception the speaker must say something like "No woman is a cunt. It's not okay to talk that way."

In a 2022 article published in *Philosophy Compass*, scholars Bianca Cepollaro, Maxime Lepoutre, and Robert Mark Simpson defined counterspeech broadly as "communication that tries to counteract potential harm brought about by other speech." It includes speech that aims to prevent toxic speech and speech that responds to it. It can take "the form of assertions, questions, imperatives, or platform-mediated acts of communication, like 'sharing.'"

In online contexts, counterspeech takes on distinct dimensions. Social media is a ripe environment for counterspeakers to call out bigotry. Online counterspeech can reach many more people than offline speech, and since the social costs and risks are different online (physical violence is less likely than in in-person confrontations), it may allow people who couldn't counterspeak in a face-to-face interaction to verbally fight derogation online. Counterspeech can warn people about the kind of bigotry they may encounter and prepare them to reject it. Counterspeaking against certain forms of implicitly harmful speech can also make bystanders less likely to accept that kind of speech when they face it in the future. Counterspeakers can offer models for responding to harmful speech that others can use, and observing that speech can offer lessons in terms of how to improve rejoinders, especially when they appear clumsy or ineffective.

When counterspeakers in an active and engaged community demonstrate they are willing to fight back against toxic speech, it can help build solidarity among the community and support for a target. In October 2020, when a reader emailed CNN reporter Nicquel Terry Ellis, "Be quiet, you house slave," she tweeted a screenshot of the message and remarked to her followers that "I don't think it gets more racist than this. But . . . I will continue to report the facts and the truth." The post received several empathetic comments, including from fellow journalists, encouraging her to stay strong, to keep reporting. Another journalist of color replied that she had received an abusive email from the same user that day. He also told her to "shut the fuck up."

There are open questions around whether counterspeech "works," complicated by the fact that not all counterspeakers share the same goals. Some people use counterspeech to promote empathy and raise consciousness. Others are focused on trying to change a person's mind or behavior. Some counterspeakers are unconcerned about modifying the views of the person perpetrating harmful speech and instead are invested in recruiting more like-minded counterspeakers to their cause. Even more, Cepollaro told me, what gets achieved or not is independent from the counterspeakers' intentions. Counterspeech can have different dimensions of success, beyond what counterspeakers would like it to do.

There are complex questions to untangle around effectiveness. Based on her review of the existing literature, Cepollaro told me the practice of repeating or amplifying toxic content can increase its spread and salience, giving it more visibility than before, more space to be discussed and assessed. Counterspeech that repeats toxic content can inadvertently invite a discussion about whether the content is true or acceptable, in some cases giving it unwarranted validity. As Cepollaro wrote in the book *Conversations Online,* a racist remark like "Go back to your country! Why did you even come here?" raises the question of who is welcome in a country, and no matter how a counterspeaker responds, their engagement can lead to more dispute. Some questions no longer need to be debated.

Skeptics of counterspeech often argue that it is emotionally and practically costly for vulnerable populations, those who tend to pay the highest price for the free speech of others. "Where inequality reigns," Tirrell wrote in her paper on the limits of counterspeech, "the odds are not in favor of someone who tries to combat the bad speech of the powerful with the more speech of the vulnerable."

I understand the desire to expose and reject harmful speech online, to strategically use more speech to undermine it. I believe it takes enormous courage. I have found that there are psychological traps here, too, especially given variability in size and makeup of a counterspeaker's audience. Counterspeakers face distinct risks and rewards on different social platforms. On Facebook, where people tend to connect with people they know, the response to a counterspeaker may be more positive and supportive. On Twitter, where relationships are more cursory and many of our profiles are open to the wider user base, counterspeakers may find less empathy and more pushback.

In 2022 I wrote a story about Senator Marsha Blackburn asking Ketanji Brown Jackson, during her Supreme Court confirmation hearing, to define *woman.* I quoted social scientists who agreed with Jackson that there is no easy answer. I was bombarded with abusive messages. When I tweeted one, with commentary, to my relatively small following to show people what I was dealing with, I received two supportive comments, while the rest excoriated me further, calling me a propagandist, telling me to get over myself, that I was so fucking stupid. One commentor noticed the unbalanced ratio of support to dissent and suggested that that should have told me something. I can't say some part of me didn't believe it did.

I have not spoken about my abuse online since.

• • •

Nancy told me that apart from the blowback to her column on Kobe Bryant's legacy, one other column and the ensuing response stand out. It was a few years earlier, in 2017, after Trump banned all refugees and suspended immigration from seven Muslim-majority countries. (Trump later dropped one country from the list.) Nancy wrote a column arguing that NFL star Tom Brady could no longer avoid questions about his support for the recently inaugurated president and that he shouldn't be allowed to.

Nancy was at the Super Bowl at that time, too. She was working out in the hotel gym when she got the message. She stopped exercising. She sat. She wept. She reported it to her boss. They had threatened her family.

"That shook me . . . that somebody would be that, I don't know if *cruel* is the right word, unfeeling, inhumane to—"

She doesn't finish.

When we stand up for ourselves, perhaps we should not always be accused of feeding the trolls. Some of us want to be loud about the threats we endure just for the right to speak. Some of us want to publicly denounce them. Many of us want the rest of the world to hear those condemnations, too.

CHAPTER 6

Trauma

WHEN I ASKED Taylor Lorenz for an interview, her first answer was no. "I don't really trust anyone to tell my story," she said over email.

This was no surprise. Taylor has been referred to as the "most harassed technology journalist in America." She has been targeted by some of the richest men on the planet, by rabid fandoms, by the far right. Tucker Carlson has belittled her in segments on Fox News. It's hard to imagine what that level of scrutiny and abuse must feel like, how careful you must be with your words, especially when handing them to someone else.

But Taylor said she was still interested in the project. She wanted to hear what I was working on. We got on the phone. Toward the end of our conversation, I mentioned that many of the women I had already interviewed had been following her story, had referred to her as an example of a woman who has experienced outrageous levels of abuse and refuses to log off.

After our call, Taylor agreed to an interview, and only she can tell you precisely why, but it was evident that despite her legitimate fears and well-earned reticence, she was still not willing to give up her voice.

Let's start where she started, with a version of a life before it was subsumed.

Taylor told me about what used to be one of the most delightful parts of her day. When she first woke, she'd eagerly tap on her favorite app, Timehop, which would distribute old photos, flashes of an ordinary life: dinner with friends, a birthday, a move to a different city, those cleansing first moments in a new apartment, capacious with possibility. That was so much fun, she'd remember. That was a good day, she'd remember. She would smile. Laugh. Sometimes she would tweet the photos. She loved this ritual, the daily doses of memory. But slowly her world changed, and those pixelated echoes began to stress the before-and-after quality of her life. Nostalgia gave way to grief. It became disturbing to think about before—the levity and openness. Harassment claimed that.

"I'm so sad for that life," she said. "My entire world has shrunk down."

Taylor, who runs the digital magazine *User Mag* after leaving her latest stint in legacy media at *The Washington Post,* built a career writing about Internet culture. She has been harassed, abused, threatened, mobbed, swatted, and stalked—online and off. She has been the victim of multiple smear campaigns. She has had those campaigns amplified by members of the mainstream media. Despite her large social media platform, she said that she still struggles to maintain a narrative of truth. When she talks about what this abuse has done to her life, her detractors accuse her of complaining.

Taylor has become a prominent figure in the fight against online violence, which has made her an even more attractive target for the right and for powerful men in tech. That visibility has also invited other forms of scrutiny. Some Black women see her as a symptom of a media environment that elevates white women's accounts of online harms at the expense of Black women's stories, which frequently go unheard. Taylor's positions at national news organizations invited public dissection of her journalistic practices, though at times some critiques have felt needlessly harsh and even exploitative. A correction on one of her stories in 2022 became a multiday news event, despite corrections being a reality of the job for almost every journalist. Sometimes conversations about Taylor's abuse veer into "well, she must have done something to deserve it." One wonders when litigating the minutiae of a targeted person's behavior can distract from the institutional and systemic failures that her mistreatment exposes.

Because for Taylor, this is not simply about people saying hateful things. "Being called a cunt is nothing to me," she said.

This is about what happens when violence cannot be contained, when the media becomes complicit in amplifying attacks. The hosts of a podcast rescinded their offer to have her on the show because they, too, were afraid. She has had speaking engagements canceled because there was no security. She said her stalker ran ads on TikTok to get her attention. She said harassers targeted her family. A hate site was created specifically to disparage her. She was scared to post pictures of her birthday party. She was scared to have a birthday party. People have told her to try to ignore it, to not live in fear. She said she almost can't believe the absurdity.

When she talked publicly about wanting to kill herself, her abusers laughed. They turned her sobs into a meme.

"I feel like I'm a frog in boiling water," she said.

A New Political Landscape

I don't think our experiences of online abuse must be the most egregious to matter. We don't need to be mobbed, or doxed, or explicitly threatened to be harmed. The acceptability of smaller-scale violences builds permissibility for larger ones. We don't need to be targeted by the most powerful men alive for our abuse to be worthy of attention. But I think it behooves us to talk about a woman experiencing abuse at this scale, about persistent and coordinated harassment, about how the media can become part of the problem. Taylor's story shows how powerful the Internet has become, the ways it has reshaped a culture.

Before she was a journalist, Taylor was an Internet personality.

She said this was somewhat accidental. After she graduated from college, she was adrift, depressed, working temp jobs. One day a girl in her cubicle introduced her to Tumblr. Taylor told me she is not being hyperbolic when she says Tumblr changed her life. She suspects it may have saved her life.

Overnight Tumblr was all Taylor wanted to do. She had never used MySpace, had unenthusiastically dabbled in Facebook, but Tumblr felt fun and creative and less toxic than other places online. At one point, she had her main Tumblr page as well as fifty other pages dedicated to her interests—photography, the singularity, sunsets, drinks, bagels, sprinkles, cloud formations, Internet jokes, funny tweets, the writer E. B. White. She learned how to program different content for different audiences, how to adjust the way she talked about something so people were likely to share it.

Taylor's proficiency on Tumblr led to job opportunities. She did social media for brands, moved on to social media at the *Daily Mail,* and eventually became director of emerging platforms at *The Hill.* She wasn't a full-time journalist yet, and while she didn't study or even aspire to journalism, she met a man around her age who was a media reporter at *The New York Times*. She thought, if this guy could be a media reporter at the *Times,* then she could be, too. She was on the Internet 24/7. People were condescending about Tumblr and YouTubers and early Internet creators. She hated that. She wanted people to understand the Internet the way she did, wanted them to understand how online influence was upending the world.

People always harassed Taylor online, and the comments were almost always sexist, but in the beginning she was generally able to dismiss them. For most of her career, the harassment had been contained. A YouTuber would make a video about her, but no one in the mainstream press covered it or cared. Most of her harassers were the YouTubers' fans. Many of them were children, whom she laughed off. Children didn't inspire terror.

"I remember the first time I got doxed, this meme account actually posted my address and phone number and everything. It was children prank-calling me, being like 'You're stupid.' Or being like 'Is this Taylor Lorenz? Haha, subscribe to PewDiePie.'" (Years later a man would open fire on two mosques in New Zealand, killing fifty-one people, and during a livestream of the massacre, the gunman was heard saying, "Subscribe to PewDiePie.")

During Gamergate, Taylor's harassment escalated, and in 2014 she saw toxic content on the Internet spread outside its most malevolent corners. She wasn't surprised when Trump won the 2016 election. She said that she saw what was happening online. She remembers the night before Trump's inauguration, covering the DeploraBall, which was organized by a pro-Trump social media organization. Taylor was surrounded by far-right Internet figures, thinking all those YouTubers were about to have so much power.

Attendees included co-organizer and right-wing provocateur Mike Cernovich, who was instrumental in inciting harassment against women during Gamergate; Jim Hoft, founder of a far-right news site known for publishing disinformation; and Gavin McInnes, co-founder of *Vice* and founder of the Proud Boys. After Trump took office, Hoft's site was granted White House press credentials. In 2017 Trump said Cernovich deserved to win a Pulitzer.

Taylor was still doing social media for *The Hill* when she headed to Charlottesville, Virginia, in August 2017 to live-stream the white supremacist Unite the Right rally. She was with the marchers as they walked through town, and she saw the gray Dodge Challenger coming. She said she stepped one foot back onto the sidewalk and watched as James Alex Fields, Jr., plowed his car into a crowd of people, killing Heather Heyer and injuring nineteen others.

It was chaos. People were screaming and bleeding. A man approached Taylor, punched her in the face, and knocked her phone out of her hands. Her live stream cut out.

Taylor didn't want her assailant to get away. She was on the ground then, and she said she reached up and gave the man a death grip. She called for the police. Some people in the crowd joined Taylor to circle him. Police eventually arrested him.

Taylor went to the police station to file a report. (Her assailant was later sentenced on a misdemeanor charge of assault and battery.) Since she was inside the station, she tweeted updates when officers brought in Fields Jr. A theory emerged online that Taylor was a deep state actor, because how else could she know so much about what was happening?

It was then, Taylor said, that she decided to become a full-time reporter. She was worried about the world, and she was worried that people didn't understand how the Internet was changing it. She started writing for *The Daily Beast*, *The Atlantic*, and then *The New York Times*.

Taylor is a target, in part, because her coverage focuses on issues the far right would rather the public not understand—how audiences are built online and how they are exploited and weaponized for profit and political gain.

As Taylor's profile continued to grow, so did the harassment. The pandemic pushed everyone online, and it pushed Taylor further online. During the pandemic, gender-based online abuse surged. Taylor watched as the Internet became everyone's "default reality." Suddenly, it wasn't just trolls or far-right figures harassing her—it was the media, too. Everything that happened on the Internet seemed worth covering, so when prominent male journalists and tech powerhouses targeted her, the media paid attention, because it was content to mine. An election had legitimatized the far right, the pandemic had made the Internet even more powerful, and many of us were watching as a woman journalist lost control of her life.

Media Manipulation

Taylor's problems, she says, aren't with individual misogynistic trolls. She indicts a much larger ecosystem that includes social media platforms and media organizations, many of which she has worked for. She targets capitalism, which she says creates the conditions for bad actors to profit from white supremacy and misogyny.

Taylor believes tech platforms need to be held accountable for their complicity and their sluggishness, but it's bigger than that, she told me. It is the content that journalists cover and the content that people online consume and amplify.

Ten years ago rape threats against women might have felt shocking, but since then strategies for abuse have grown more sophisticated, and they are aided by systems built to profit off morally outrageous content and by journalists who are under enormous pressure to cover whatever is trending on the Internet. Claire Wardle, co-founder and co-director of the Information Futures Lab at Brown University, told me that coordinated harassment campaigns that seek mainstream attention are "a much more strategic use of power over women than 'Oh, I'm going to rape you.'"

The danger, she told me, lies in the amplification provided by the mainstream media. "The media is what gives you the audience," she said. "It's only effective if the abuse moves from 4chan all the way through to Fox News."

Scholar Alice Marwick said that while some of the spread happens organically, there is an entire cottage industry that attempts to "trade up the chain," to get enough smaller pockets of the Internet talking about something that it can move its way into more mainstream publications. Marwick said this is common in the far-right media ecosystem, including with outlets like Fox News and *The Daily Caller*.

"They only really exist as controversies because somebody is propping them up as a controversy," Marwick said, "and they get picked up by one of these blogs, and then Fox News picks it up because they need a constant stream of content that fuels their audience's outrage, because that's their entire business model."

In March 2021, after Taylor had tweeted that online harassment had "destroyed my life," Tucker Carlson reported on it for a segment on *Tucker Carlson Tonight:* "Destroyed her life, really? By most people's standards, Taylor Lorenz would seem to have a pretty good life, one of the best lives in the country, in fact. Lots of people are suffering right now, but no one is suffering quite as much as Taylor Lorenz."

Harmful narratives targeting Taylor are found across a broad swath of media. In 2022 *New York* magazine wrote a feature on Taylor joining *The Washington Post,* characterizing her as juvenile and petty, referring to her "giggling" during the interview, as if she were a child. A *New York Post* op-ed called Taylor a "Mean Girl basket [case]." A *Gawker* story once wrote about her under the headline "Welcome to the Little Bitch Olympics."

When Taylor reported on the woman running the powerful right-wing Twitter account Libs of TikTok, which is rife with racist, transphobic, and anti-LGBTQ posts, she was pilloried by the right on social media, and the "controversy" was picked up by conservative news outlets. Taylor knocked on doors in an attempt to speak with the woman running Libs of TikTok, and a tweet from the account showed a photo of Taylor outside her door, writing: "Which of my relatives did you enjoy harassing the most at their homes yesterday?" In an op-ed, media watchdog Poynter stressed that knocking on doors to get comment from a person you are writing about is standard practice in journalism, yet a headline on Fox News read: "Washington Post's Taylor Lorenz doxxes Libs of TikTok days after decrying online harassment of women."

Most of us may not easily draw parallels between the behavior of trolls and the behavior of corporate media, but as scholar Whitney Phillips writes in *This Is Why We Can't Have Nice Things,* the line between the two is "thin and at times nonexistent," and the "primary difference is that, for trolls, exploitation is a leisure activity. For corporate media, it's a business strategy."

Dissociation, Desensitization, Numbing

How do you cope with losing a version of your life?

When women are abused online, the demand may be silence, but the impact is often trauma, which rips apart our narratives about who we are, how other people are, what life means, and how the world works.

"The reason that these strategies are working is because of how severe the impact is on mental and physical health," PEN America's Viktorya Vilk told me. "It's psychological warfare. It's effective."

Taylor was able to cope with misogyny in tech. She dismissed sexist attacks and told herself she was competent. She coped with far-right YouTubers, even ones with significant followings, making videos about her, finding humor in an eleven-year-old's death threat.

But the mainstream media's amplification of far-right attacks, she said, set off different reactions. "It's a different type of state. It's not just like I'm ignoring it. It's like I cannot feel emotion," she said. "I've gone through so much that . . . I feel numb."

Many women I interviewed said that when they reflected on their experiences of intense online abuse, they were "numb" or "dissociating." They described experiences of complete overwhelm. Activist Jaclyn Friedman said when she was in the worst of her abuse, she had the sensation of leaving her body. Disinformation researcher Abbie Richards said there were times she was so overcome she blacked out. Philosopher Kate Manne said she has almost no memory of the worst of her attacks. When she was targeted by psychologist and conservative media commentator Jordan Peterson, Manne said she had to take a Xanax and sleep for twenty-four hours "because the stress was so enormous."

Women who used the terms *numb* and *dissociating* had often experienced one or more large-scale online attacks. Emotional numbing and dissociation are things our brains typically do unconsciously to protect us when our systems become overwhelmed. These are not necessarily coping strategies, but the frequency with which I heard these terms used suggests they are relevant to the question of how women are managing their experiences. Not every way we survive involves conscious choice.

In the absence of a clinical evaluation, it's impossible to know precisely what psychological phenomena were at work in these women's cases. All I have are women's memories and the language they find most accurate and accessible to describe them. Some women likely were dissociating, especially when their online harassment was layered on top of or triggered an earlier trauma. But it is also possible, psychologists told me, that women who said they dissociated were referring not to dissociation in clinical terms but to a combination of emotional numbing and cognitive avoidance, which can involve distraction and thought suppression. All these responses can be useful, and they can also have harmful effects.

Emotional numbness is characterized by diminished interest, disconnection, and lack of emotional response. It can feel like emptiness or a daze. It can frequently be associated with trauma. Taylor has been diagnosed with PTSD.

When you are emotionally numb, you are not constantly in a state of alert. You don't feel miserable, don't feel the awful things happening to you, but you may not feel good things either. You may not care about things that once mattered to you. You may stop caring about yourself.

Elana Newman, research director for the Dart Center for Journalism and Trauma, said that emotional numbing can overlap with dissociation. Dissociation involves an alteration of a conscious experience, a breakdown in functioning. It generally happens involuntarily during an experience of something truly overwhelming, an immediate threat to life, to bodily integrity, to the self. Dissociation is protective, to spare a person from a truly intolerable experience. One psychologist described it to me as a last-ditch emergency mechanism, because it also means losing control and access to important information for the sake of protection from pain.

Katherine Porterfield, a clinical psychologist who specializes in trauma and is a consulting psychologist at the Bellevue Hospital Program for Survivors of Torture, said that dissociation is an adaptation to trauma, but it becomes a symptom when it persists in the aftermath of trauma.

While a traditional conception of dissociation involves situations where the disconnect is necessary for preventing immediate pain or harm but the recurrence of the reaction becomes maladaptive, some clinicians are broadening their thinking on dissociation to include a wider continuum of involuntary reactions that involve losing touch with one's immediate surroundings. Clinical trauma specialist Jamie Marich wrote that the goal of dissociation is "to protect oneself and to get one's needs met." Going away from your own mind can protect you from distress, she said, whether rooted in pain or boredom, and we don't need to have a dissociative disorder to have engaged in dissociative behaviors, whether we're daydreaming or doomscrolling.

The other term Taylor and several women I interviewed used when speaking about their online abuse was *desensitization*. They said they had become desensitized to their attacks, which is why they believe they've been able to remain online.

Without context, *desensitization* is a neutral term that refers to diminished responsiveness. The reduced responses can be physiological, emotional, cognitive, or behavioral. In trauma work, desensitization is something a clinician can aim for with a patient, to reduce their reactivity to something that they perceive as dangerous but is not, or at least not as dangerous as they think. It can be used to work with a patient who was in a car accident and is now struggling to get back inside their vehicle.

But as psychologist Emily Sachs told me, a person wouldn't want to reduce reactivity to something that is dangerous. While it may be useful for a woman to be less emotionally responsive to being called a bitch and a cunt online—she can keep posting, writing, speaking, with less disruption to her life—that diminished responsiveness can normalize violence and likely also exact other psychological costs.

When women spoke with me about being desensitized to online abuse, I felt they viewed it as necessary for their continued participation in digital spaces. It may be. But just because someone becomes desensitized to something doesn't mean that they are not experiencing other kinds of distress. Literature on living in violent communities suggests it does desensitize you to violence, but people in those communities are still anxious and depressed and experience trauma symptoms. Maybe you're desensitized to your abuse, but that doesn't mean it isn't adversely impacting you. It can lend itself to other forms of disillusionment and withdrawal. It can chip away at your resolve. It can lead to self-censoring. Maybe you start to lose the point of what you were saying. Maybe you lose the pride in what you have been doing. Maybe you are not doing what you came here to do, at least not in the way you wanted. There's a subtle chain of loss in optimism and identity.

Desensitization can also compromise our ability to evaluate whether something is a credible threat. After years of abuse, Dr. Leah Torres told me she believes desensitization has made her threat response system less effective. She once got a threatening message that she discussed with a friend, and they encouraged her to report it to the FBI. She was surprised. What her friend saw as a credible threat, Leah viewed as a routine and dismissible message.

"In order to combat violence," Leah said, "you have to fight your own experiences."

• • •

In 2021, the International Women's Media Foundation (IWMF) released a statement condemning the attacks against Taylor, writing that it was "appalled by the relentless online smear campaign" and stressing that "women journalists must be able to do their jobs without fear for their lives."

Taylor has never stopped reporting, writing, or speaking, despite the abuse.

"Outside of her beat, outside of the cultural reporting that she does, the technology reporting that she does, she will respond and hold up a mirror to these types of campaigns that are trying to strip her of her work, strip her of her platform," IWMF director of communications Charlotte Fox told me. "The aggressors don't like someone fighting back, and their ultimate goal is to keep them in this perpetual cycle of trauma."

Most of the time, Taylor said she finds herself vacillating between numbness and rage. In 2020, when her abuse really began to take off, fellow journalists would tweet in support. It was not unlike the people in Charlottesville who helped her circle her attacker. Eventually, the bulk of that support dissolved, possibly because those standing with her became targets themselves. The harassment, however, has never stopped.

Women experiencing abuse can either completely log off—a defeating prospect for many and an impossibility for others—or they can continue to cope, which can sometimes mean detaching from themselves, from their bodies, from other people. Those bearing witness, however, have the luxury of looking away. As trauma expert Dr. Judith Herman writes, amnesia can also be social: "In the absence of strong political movements for human rights, the active process of bearing witness inevitably gives way to the active process of forgetting. Repression, dissociation, and denial are phenomena of social as well as individual consciousness."

Every time Taylor sees her mother, her mother says, "Taylor, you gotta quit." But Taylor says she won't. Or can't. Partly because it feels unjust, but also because she can't picture her life without the Internet. It has given her community, a purpose, and a platform. She says it's almost impossible to imagine not living an extremely online life.

Taylor says she's determined to keep going, to keep writing, to keep reporting on the challenges for women online.

But she also wonders: "When are they going to take this shit seriously?"

I suppose she's talking about all of us.

CHAPTER 7

Information Disorder

IN A BOOK ABOUT VIOLENCE, a love story broke through.

Céline Gounder and Grant Wahl met when they were both students at Princeton. They were young and ambitious, earnest in the way so many of us are at the beginning of our lives. Both wrote for the school's paper, *The Daily Princetonian*. One of the first things Céline remembers Grant saying to her was "I read your byline."

Céline was studying pre-med and Grant was studying politics and honing his skills as a sportswriter. While Céline took no interest in sports, and Grant knew little about biology, they shared a certain curiosity about the world. Part of what Grant thought was so interesting about covering soccer, in particular, was that as a global sport, it was a way of understanding a local culture, a politics, a history. Part of what Céline thought was so interesting about public health was the way it demands you understand a culture, a politics, a history.

Six years later Céline knew she wanted to marry Grant. She planned to propose at a lavish dinner, but before the meal Grant's menu curled into a candle and caught fire. The moment was not right. When they returned to their studio apartment on Manhattan's Upper West Side, Céline put the silver ring with the inscription "Marry me?" in her underwear drawer for safekeeping. But when she woke the next morning, she decided there was no right moment, that there was only ever this moment, so she directed Grant to her drawer. He said yes.

There's a picture of them from that year. They're dancing. She is smiling and looking away. He is looking at her.

Céline became a doctor, an expert in infectious disease who would help guide a nation during a global pandemic. Grant became a celebrated sports journalist who used his platform to advocate for women's soccer and human rights. They loved each other for more than a quarter of a century.

In 2022 Grant traveled to Qatar to cover his eighth World Cup. In the closing minutes of a match between Argentina and the Netherlands, Grant collapsed in the press box. Emergency service workers performed chest compressions before taking him out of the stadium. Grant died at forty-nine years old.

When Céline learned of his death, she felt all the things one feels when someone they spent their life with is suddenly gone. She also felt urgency. Backlash against public health and vaccination was mounting, and she knew what was coming. Even in her grief she knew she needed to act.

As soon as news about Grant's death became public, false claims spread. Céline watched as strangers blamed Grant's death on the COVID vaccines she was working to get into the bodies of the American public. She watched as people tried to link Grant's death to myocarditis, an inflammation of the heart muscle and a rare side effect of the COVID vaccine. COVID-19 poses a higher risk for myocarditis than the vaccine itself.

Online people told her: "Now you understand that you killed your poor husband."

Céline was clinical. She put anguish away. After she told family that Grant had died, her first call was to the White House. The plan was to fly Grant's body home to perform an autopsy. She wanted facts to fight lies. Grant's legacy was hers to protect.

Four days later she got the autopsy results: a slow-growing, undetected aortic aneurysm. She released a statement with the findings. She did interviews with *The New York Times*, CBS, and NPR.

As a public health professional during a politically polarizing pandemic, Céline experienced frequent online abuse. She avoided what she could. She never responded. She didn't check Twitter mentions. She was generally unmoved by rape threats. "Idiots," she thought. But she could not dismiss what people said about Grant. When people online blamed her for his death, she told me she felt like she wanted to vomit.

After Grant died, many people reached out to express condolences, to help get Grant's body home, to discuss the memorial. Céline couldn't ignore her messages. It was impossible to screen out the hate, because so much of it was mixed in with support and love.

Céline needed to grieve, but the harassment was relentless. She divided herself. She put Grant in one place, so she could function in another.

Grant's memorial was on December 21, 2022. But just after the new year, Buffalo Bills safety Damar Hamlin went into cardiac arrest during a game against the Cincinnati Bengals, collapsing on the field, and the anti-vaccination community again grew feverish. Céline's harassment ramped up. Again, she divided herself.

Céline wanted to reframe the narrative. In an op-ed for *The New York Times,* she wrote: "When disinformation profiteers leverage tragedies like Grant's and Mr. Hamlin's for their personal gain, they retraumatize families, compromise our ability to interpret information and distinguish truth from lies and put all of us at risk. The results of allowing this to continue will be disastrous."

A couple of weeks after the piece ran, Céline's harassment finally slowed. She unpacked her anguish. She began the endless process of making sense of a world without Grant, of trying to keep the public healthy despite knowing they would never stop reminding her of what she lost.

Disinformation, Misinformation, Malinformation

Rarely is the problem of disinformation framed in connection with the problem of online abuse. Put crudely, one is considered a problem of incorrect facts, the other a problem of abusive behavior. But when social media companies allow disinformation to spread, it can become weaponized in online harassment campaigns, like the one directed at Céline. Disinformation, which can be spread by humans and by bots, can also be used to create more insidious harms, especially against marginalized communities. As technology analyst Shireen Mitchell told *The Washington Post* in 2020, agents of disinformation pretending to be Black women online, itself an abusive behavior, contributed to digital voter suppression in the 2016 election. During the 2024 presidential election, a CNN investigation in collaboration with the Centre for Information Resilience (CIR) found that women's online photos were stolen by unknown actors and used on social media to promote Trump and his running mate, Senator J. D. Vance of Ohio. Deepfake technology is being used to convince people that women have done or said things they never did. A 2023 report on the state of deepfakes found that 98 percent of all deepfake videos online are pornographic and 99 percent of those targeted are women. We cannot think of online abuse as only death threats and slurs but as a broader arrangement of tactics and behaviors and the dissemination of falsehoods that contribute to a culture where violence against women is both perfunctory and conspiratorial.

Disinformation online is often used to create narratives about women that draw on sexist, misogynistic, and racist stereotypes. To channel anger toward women, people have to believe incorrect facts about women—that we lie about rape, that we are gold diggers, that we don't belong in positions of power and are too emotional to be competent. Disinformation is used to justify online attacks.

Claire Wardle, co-founder and co-director of the Information

Futures Lab at Brown University, told me that while many people use the terms *disinformation* and *misinformation* interchangeably, they are distinct. Disinformation is content that is intentionally false. People can create and share this information to make money, to achieve political goals, or to create harm. Disinformation becomes misinformation when people share it without knowing it's false. A third form of information in the ecosystem that gets less attention is called *malinformation*, content that is based on reality but is often used maliciously and out of context (think revenge porn). Disinformation, misinformation, and malinformation are part of the problem Wardle calls "information disorder," a term for the spectrum of content polluting the Internet and media more broadly.

Gender disinformation is a newer phrase that some individuals and organizations use to denote information that is intended to cause harm to women or, more broadly, to people of diverse genders and sexualities. The U.S. State Department defines it as using "false or misleading gender- and sex-based narratives, often with some degree of coordination, to deter women from participating in the public sphere."

In 2007 activist Jaclyn Friedman published a piece about victim blaming that included a reference to her own sexual assault. Afterward, she was inundated with skepticism and hate. Some of her harassers claimed they knew she was lying because they had looked up her police report. Friedman said she had never filed a police report.

Zoë Quinn, who was at the center of Gamergate, became a target after an ex-boyfriend posted an online manifesto accusing Quinn of sleeping with a video game reviewer in exchange for a positive review of Quinn's game. This was a falsehood. The purported review doesn't exist.

When Taylor Lorenz knocked on doors to get a comment from the woman running the right-wing Twitter account Libs of TikTok, conservative news outlets accused her of doxing a private citizen, even though knocking on doors is a common practice in journalism, both to confirm the identity of the person a journalist is writing about and to give the subject of a story an opportunity to be quoted.

When Representative Alexandria Ocasio-Cortez went on Instagram Live in 2021 to discuss how the trauma of the January 6 insurrection had compounded a previous experience of sexual assault, social media posts went viral with claims that she lied about facing rioters in the Capitol Building. But in her Instagram Live, Ocasio-Cortez clearly stated she was in her office near the Capitol and that she feared for her life before discovering the person banging on her office door was a Capitol police officer.

The abuse directed at Céline, a high-profile and high-achieving woman of color, was fueled by vaccine disinformation, which became widespread misinformation, but it was also rife with the sexist, racist language that is present in many gender disinformation campaigns. People said Céline was a "liar" who was profiting from the drug companies. People called her "a sick bitch." They said they did not want her "crooked opinion," hearkening back to "crooked Hillary," what Donald Trump called Hillary Clinton during the 2016 presidential race. One Twitter user was so disgusted by her CBS interview on vaccines, they told her to "RESIGN & GO WASH BATHROOMS."

• • •

Technology companies are rightly criticized for profiting from disinformation. A 2020 report from the Center for Countering Digital Hate found that social media platforms didn't act on 95 percent of the anti-vaccine misinformation reported to them.

But while social media companies' failures to address dis- and misinformation make headlines, far less attention is paid to how these problems predate the Internet.

Scholar Alice Marwick said that her research center, the Center for Information, Technology, and Public Life, explores a longer history of propaganda in the United States. It examines the ways that not only social media platforms but also legacy media spread disinformation. One of its case studies features the *welfare queen*, a term that has been around since the 1970s. Marwick and her co-authors wrote that it is "rooted in racial, classed, and gendered stereotypes." The trope has been used to stigmatize public benefits and to justify "the reduction of welfare in later policy reform."

Marwick said that disinformation is not just incorrect facts on the Internet but also narratives that are generated by the dominant culture about marginalized and minoritized communities.

"If you think about the idea that Black people are more inclined to criminality—something that fueled hundreds and hundreds of years of racism, Jim Crow, the carceral apparatus, prison expansion—all these things are underpinned by these racist lies that justify the actions of the dominant population," Marwick explained.

The reason so many contemporary disinformation narratives work, she said, is that in many cases they are building on preexisting racist, sexist narratives that have circulated in the culture for generations. When disinformation online is used to spread the false belief that drag queen story hours are dangerous for children, it's building on decades of homophobic disinformation claiming that queer people are a danger to kids. When disinformation is used to attack the character and credibility of women online, it is effective because it is based on age-old sexist narratives. When Zoë Quinn was accused of sleeping with a video game writer in exchange for a good review, people found it plausible because they have been primed to believe that women sleep their way to the top.* During the 2020 presidential race, Kamala Harris was accused of the same. At a rally in the last days of his 2024 presidential campaign, Trump laughed after a rally-goer implied Harris had worked as a prostitute.

* Zoë Quinn now uses they/them pronouns.

"We don't even see this as disinformation," Marwick said. "We see these as stereotypes about women, but they are stories about women's inherent being that are constantly reinterpreted and re-broadcast in every single form of media and conversation that you can think of, from dumb blonde jokes to true crime narratives."

Marwick said rather than thinking of disinformation as incorrect facts, she and her center think of them as sets of stories, narratives that often get retold in harassment campaigns, especially regarding people of color and queer folks.

Agents of disinformation have for years targeted women and people of color online. A 2018 report from Stop Online Violence Against Women, authored by Shireen Mitchell, found that of the 3,500 ads that the Russian Internet Research Agency posted on Facebook during the 2016 election, the majority targeted "Black Identity and culture . . . with the intent to engage in voter suppression of Black voters." Disinformation has been used to spread misleading vaccine information to Black and Latinx communities and to drive a wedge between them, which is precisely the kind of distraction that keeps people from addressing the true sources of their suffering. For many people of color, experiences of discrimination have contributed to mistrust of medicine and medical information. Black Americans have long experienced abuse and exploitation in the medical system—Black bodies pulled from graves for scientific study, Black women sterilized without their consent, and the notorious Tuskegee syphilis experiment, when the U.S. government let Black men with syphilis go untreated so it could observe the damage the disease did to their bodies, even after penicillin was discovered as a cure.

When researchers work to expose disinformation, they often become targets themselves, and many of them do not have institutional support. The Black feminists who exposed disinformation with the hashtag #YourSlipIsShowing were working independently, not affiliated with a prestigious think tank or center, and when they were threatened online, there was no institution to stand with them.

Women in politics, and specifically women of color, are especially vulnerable to becoming targets of disinformation campaigns. A report from the Center for Democracy and Technology found that during the 2020 election, women of color candidates were twice as likely as other candidates to be targeted with, or to be the subject of, mis- and disinformation, and they were the most likely to be the subject of posts that combined mis- and disinformation and abuse. Research from the organization #ShePersisted found that gendered disinformation campaigns target not only women in politics but "their families also, with rape threats against their young children becoming an ever more common and deeply disturbing phenomenon."

In 2024, the Center for Countering Digital Hate collected more than five hundred thousand comments on Instagram posts from leading Republican and Democrat women politicians running for office. The Democratic women included Vice President Kamala Harris and Representatives Alexandria Ocasio-Cortez and Jasmine Crockett, while the Republican women included Representatives Marjorie Taylor Greene and Lauren Boebert. CCDH used a machine-learning tool to gauge the toxicity of comments, ultimately reporting one thousand comments that their researchers felt were definitively in breach of Instagram's community guidelines on abuse, including sexist abuse such as "Make Rape Legal," racist abuse such as "We don't want blacks around us no matter who they are," and threatening comments such as "Death to her and her supporters." Of the one thousand comments CCDH reported, 93 percent remained after one week.

A 2019 *Guardian* investigation found that an Israeli-based group targeted Muslim lawmakers Ilhan Omar of Minnesota and Rashida Tlaib of Michigan, who that year had become the first Muslim women to serve in Congress and who have since experienced Islamophobia during their time in office. *The Guardian* found that the group "co-opted at least 21 organically grown far-right pages" to "stoke deep hatred of Islam across the western world and influence politics in Australia, Canada, the UK and the US."

Disinformation online continues a long history of propaganda by individuals and groups in pursuit of political and economic goals, and now Big Tech directly profits. A 2022 ProPublica analysis found that Google, despite publicly committing to fighting disinformation, "placed ads from major brands on global websites that spread false claims on such topics as vaccines, COVID-19, climate change and elections."

Compartmentalization

Céline sent me dozens of abusive messages that she had received in the year since Grant's death. Most were not influencers with large platforms but appeared to be everyday people who were exposed to lies, believed them, and directed their fear and anger toward Céline.

Information disorder creates confusion. It can sometimes be difficult to know who the bad actor is, who is gullible, what is fact, what is fiction, and what is true but out of context. Information disorder creates chaos in individual lives and in public life, in politics and in communities. Céline coped with the chaos by cultivating an internal organization. She told me the only way she survived Grant's death and the abuse was by compartmentalizing, which the American Psychological Association defines as a process where "thoughts and feelings that seem to conflict or to be incompatible are isolated from each other in separate and apparently impermeable psychic compartments."

Grant was gone, and people were calling Céline a "Satanist," wondering how she could look at herself in the mirror. They called her a "fake news maggot," a "criminal bitch," an "ugly piece of shit on the inside" who looked "ten years older than you are" on the outside. They told her they wished her "dying day" would come soon.

A woman left her a voicemail, calling her a widow by choice. A man with a UCLA email address told her: "People like you are going to swing from lampposts." Two months after Grant's death, Céline did an interview with CBS on COVID vaccines, debunking false reports that they cause "sudden death" via cardiac complications. Afterward, some messages were so terrifying that she reported them to the FBI.

But she still had to work, to breathe and think, to speak and move.

When information comes through that is difficult to process—a tragic loss, a threat to your life, a racist remark—you can experience an unwanted emotional reaction that you then try to override, neuropsychologist Negar Fani told me, "cognitively doing a lot of work to pack it away so that you can function, so you can get your work done, get your tasks done. This is the work that our brain does in a sophisticated way, but it takes a lot of resources."

When I spoke with Céline, I could feel just how much she had tucked inside. Journalists are practiced in asking the same question differently, trying to get to the heart of a matter. Céline wouldn't let me near her heart. During our interviews, she talked of Grant's death and her abuse clinically. She didn't waver or need a moment to catch her breath. She never got flustered.

Céline is a doctor. She is practiced in compartmentalizing, a skill useful for people in the medical profession. She told me one family member remarked that her ability to compartmentalize was "Herculean," so when her husband died across the world and people blamed her for it, she used this skill to survive grief. She divided herself so she did not lose herself. But she admits that when it came to the abuse around Grant, "there is a very real stress to that. How do you bottle that up? How do you contain it there? There's real effort involved."

Compartmentalization is often described as a psychological defense mechanism. It can be an unconscious reflex, or a conscious decision. It is often the latter for Céline. It can involve diversions, setting something aside, refocusing attention so that you can manage your feelings. In the *Encyclopedia of Personality and Individual Differences,* researchers Vera Békés, Yocheved Ayden Ferstenberg, and J. Christopher Perry write that compartmentalization is when a person "separates various aspects of the self (for example, beliefs, social roles, emotions, cognitions) and can only access one of these aspects at a given time." This kind of separation allows a person "to have conflicting ideas or self-concepts without experiencing tension" around the contradictions that naturally arise. Compartmentalization is sometimes referred to as a form of dissociation.

In practice, compartmentalization can look like saying to yourself, "I'm going to do something that takes over my body's sensation experience, affect experience, intellectual experience in order to distract from that thing that felt harmful and difficult," psychotherapist Abra Poindexter explained to me. "We all need to actually do that. . . . You have an argument with your partner in the morning, but you still have to go do something. You have to compartmentalize to some degree."

Many professions demand it. Doctors who treat sick patients. Journalists who cover violence. Therapists who treat suicidal people. Researchers who study hate and extremism.

As Poindexter says, compartmentalization can be an effective way to switch activities, when you need to deal with something distressing later. You can use it to minimize painful emotions so they do not interfere with something else you want or need to focus on. It is sometimes used to organize positive and negative beliefs about the self. It is least adaptive when it involves splitting (failing to reconcile positive and negative attributes into a whole understanding), repression (blocking undesirable ideas), and dissociation defenses. This combination can allow someone to separate parts of their life that conflict with their usual identity. Békés, Ferstenberg, and Perry give the example of a person who publicly preaches against pornography but keeps a robust collection of it at home.

While writing this book, I found compartmentalization to be a crucial defensive skill. The content of this project and the demands of living were sometimes incompatible. Many times I had to put aside my own life to focus careful attention on a subject. At the end of the day, I had to lay down my work, my frustration, and my fear to listen well to my children. These were imperfect acts. I have recordings of interviews where the girls burst in. I had days when my own pain made its way into the transcripts.

When I reflect on how I coped with my online abuse, I realize how essential compartmentalization has always been, how it has helped me find pockets of safety and peace, productivity and joy.

I compartmentalized with my reader who told me I was not a person. I was holding my baby, and I would not have those words in that room in that moment. Eventually, I returned to them, examining what those words necessitated. It was after that revisiting that I began in earnest to write this book.

The key, therapists say, is to come back to it. Poindexter said a person needs to face the impact of an experience to help themselves through it. It's important to ask yourself: *How did that make me feel? How did it make my body feel? How can I deal with this? What do I need that I did not get when this happened, and how can I give this to myself now?*

"We can only try to outrun it for so long," she said.

• • •

Occasionally, Céline will pause long enough that the grief demands noticing, though she admits she is still reticent to be entirely present with her loss.

"I think I have not, still now, fully attended to it," she told me several months after Grant's death, during the 2023 Women's World Cup. She told me Grant had wanted his next book to be about the women's team.

Céline was still pouring herself into her work, a passion she and Grant had shared.

"I don't think he would be surprised," she said of Grant, to know that "I've continued just to be who I am."

Céline told me she and Grant used to read *The New York Times* together every morning. She misses that. She wrote her *New York Times* op-ed on disinformation to educate the public, but she also wrote it to honor her husband, that young man from Princeton, so curious about the world, who loved soccer, who loved her, and who most certainly would have loved to see her byline.

CHAPTER 8

Online, Offline, What Line?

THIS CHAPTER IS a departure in form. To understand what is happening on the Internet, we need to also look at what is happening off the Internet. We begin with a woman who was fluent in the language of misogyny long before she had a social media account.

The first moment Brooke Nichols could remember wanting to be something, she wanted to be a welder.

When she was twelve, Brooke was diagnosed with Tourette's syndrome and obsessive-compulsive disorder. Her struggles in the classroom led to a placement in a special education program that trained kids for the workforce. Adults in the program pushed her toward horticulture, toward nursing, toward cosmetology. Brooke wasn't interested in any of it.

The adults told her she was ridiculous. What girl wanted to weld? Brooke found herself stuck in a chasm between makeup and metal. But she was indefatigable, and eventually the adults relented. Brooke was the only girl in the welding program. The boys rolled their eyes. They said she was too pretty to weld. Her teacher noticed she wore mascara and said maybe she really should consider cosmetology. Brooke didn't understand what mascara had to do with it.

At first, no one would let her touch the tools. Her instructors wouldn't let her lift steel, which they said was too heavy. Instead, she watched and listened. She was painstakingly attentive, so when they finally let her weld, she became very good very fast. She was a "sick welder," she told me. But then she had different problems, because not only was she a girl who didn't belong, but she was skilled and was considered a threat. The boys joked that the only reason she was a better welder than they were was because of her tits.

When she was seventeen, she competed in a New York state competition. While she was there, no one spoke to her. No one helped her set up her machine. She placed higher than most of the boys.

Sometimes Brooke tried to talk about how hard it was, but no one really understood. She tried to talk to her first welding teacher. Eventually, he took her under his wing, but still he told her: "Just ignore it."

"How?" she wondered.

When she turned eighteen, she got her first welding job. That's when intense sexual harassment started. The boys in her class had made comments before, but she said it felt more like being "roasted." This was different. She was a teenager, and she was among men. A man told her he'd had a dream about her naked. She reported it to the only other woman on the jobsite, who worked in the office.

The woman told Brooke that if they touched her, to come tell her, but otherwise:

"Just ignore them."

Brooke reached a point when she couldn't take it anymore, so she stopped welding. As scholar Sara Ahmed writes, "Being able to leave requires material resources, but it also requires an act of will, of not being willing to do something when it compromises your ability to be something."

Brooke started a cleaning service. When she was nineteen, she became pregnant with her first daughter, and she kept cleaning until she was eight months along, but then her customers started telling her how uncomfortable they felt watching her on her hands and knees. Her husband was a contractor running his own painting company, so she decided to quit cleaning and try on "stay-at-home mom."

She entered the world of mommy groups, though that wasn't quite a fit. She was always the youngest mom in the bunch. A teen mom, really, and they never let her forget it. But despite her discomfort, the time she spent in those groups would inspire her first entrepreneurial venture. Brooke, who had been using cloth diapers on her daughter, heard some moms wishing for a cloth diaper delivery service, so she decided to start one. A year later she had her second daughter, and over the next decade, she said, her diaper company serviced thousands of families.

"I was really proud of what I did. I had nothing. I had no college education. In special education, they did the best they could, but they didn't teach you well. I can't do fifth-grade math. I can't even help my kids with their homework. You know what I mean?"

Brooke had her third daughter when she was twenty-seven. But after eight years of running her diaper business, she felt she had hit a ceiling. Truthfully, she told me, she itched to get back into the trades. She hated the way men treated her, but she loved to build. She loved mastering a skill, and she knew that the more skilled she became, the more money she'd make. Her daughters were getting older, their worlds were getting bigger, and she wanted them to have access to opportunities that she did not.

When Brooke was twenty-nine, she returned to the trades. She began to weld again. She didn't ignore the men's insults, their derision, their taunts. She didn't ignore when they said terrible things to her, when they told her she was being "a pussy."

"Don't talk to me like that," she scolded.

They would play dumb. They would say she was too sensitive. They would mess with her work, mess with her machine. One time Brooke lost it. She yelled and cried. The owner saw and told the men they needed to stop, and most of them did, because he had power.

We're sitting together, and I see her hands are shaking while she tells this part of the story.

She told me that one day her boss asked her to work late with a male co-worker. Brooke didn't trust the man, but she needed the overtime. She worked the shift, and she was in the front office when the man called her to the back, to the barn.

She knew what was ahead, but she didn't feel like she could stop walking.

When she got to the barn, he called her "toots" and said he needed help. He immediately hit on her. He asked questions. He asked, What does your husband do?

"I knew. I'm like 'I'm in fucking trouble. I'm in a barn. There's no way for me to get out.' . . . He's just like 'Oh, you going to make this hard or you going to make this easy?' "

She thought of her girls. She didn't fight or freeze or flee. She fawned. She flirted. She told him how blue his eyes were. She gave him a fake phone number. She said Friday, that's when they would meet. Not tonight, but Friday at the bar. He bought it.

When she heard the welder go on, she knew he was distracted so she ran. She bolted back to the office, locked the door. She called her boss, who fired the worker. But he came back to get his last paycheck. She was there that day.

"Fucking bitch," he told her.

Artificial Boundaries

Online and offline violence may be framed as separate cruelties, but they are part of a single story.

In the earliest days of the Internet, we were told that cyberspace was distinct from our real lives and even our real selves, a place where, as scholar Adrienne Massanari told me, we could embrace "this matrix experience that was separate from our everyday embodied experience."

Except it wasn't true. "There's nothing 'online' or 'offline,'" Massanari told me. "It's all part of the same world."

The language that women hear online is not confined to the Internet, especially for women who navigate predominantly male spaces. When I watched Julie DiCaro and Sarah Spain's moving PSA on online abuse, I was not entirely convinced by the text that appears on screen at the end: "We wouldn't say it to their faces. So let's not type it."

It's true the Internet is different. When you can pretend you are not human, when you can pretend your target is not human, when you can say horrific things without any real cost or consequence, this is a different environment with different norms of behavior. People online are viler, louder, and their written words take on a vicious permanence. Still, it's not as if these are the only places we hear these words. Some people wouldn't say it to our faces, but many people would and do. They talk to us like we're nothing. They talk to us this way in every place we are.

After work, in bed with my kid, I open my email: "Cunt."

At work, eating chips at my desk, I open my Facebook Messenger: "Truly disgusting."

Before work, on the subway, I open Twitter: "Dear God, shut up."

In the car, on the treadmill, in the kitchen, at the bar, I open my phone: "Brainless." "Liar." "Shameful." "Simpleton." "Empty." "Soulless." "Pig."

But also.

When I was eighteen, in the frat house, a man spat: "Crazy bitch."

When I was twenty, in a club, a man remarked to his friend: "She dances like a whore."

When I was thirty-six, on a train, a suit told me: "Your voice! So annoying!"

I was not on the Internet when a man I didn't know put his hand under my skirt. When I yelled, he said: "Don't be such a cunt."

I was not on the Internet but on a couch, when I drank too much, when I couldn't move, and a man started to put his hand down my pants. "Shhhh," he whispered.

Always, silence is the demand.

Remember.

"Fucking bitch," Representative Ted Yoho allegedly called Representative Alexandria Ocasio-Cortez on the steps of the U.S. Capitol.

Remember.

A broadcaster told a former Australian prime minister to "shove a sock" down former New Zealand prime minister Jacinda Ardern's throat.

Remember the words Donald Trump used to describe us: "fat," "pig," "dog," "slob," "disgusting, both inside and out." Remember when he said Kamala Harris "happened to turn Black," when he publicly called her "nasty," when it was reported he privately called her a "bitch."

When Alexandria Onuoha was called a "black bitch" online for publishing an op-ed about white supremacy in her dance program, it was not the first time she had heard those words. They had been said to her face a year before, by a stranger on a train platform.

We have to reject the notion that this is just about the Internet, that we can just log off, that it isn't real.

Sareeta Amrute, an anthropologist at the New School who studies race, labor, and class in global tech economies, told me that when she conducts ethnographic interviews, it's apparent there is continuity between what is happening for women online and what is happening for them offline. She has talked to women from the United States, India, the U.K., and Australia, and she said: "For women, especially women from intersecting marginalized identities, the online space doesn't feel new. It feels like a continuation in a new medium of what they've been experiencing all along."

This story about violence reaches back to the time before smartphones and virtual mobs, implicating a system of power that deemed women inferior, that readily punishes us for behaving as though we are not. Writer Soraya Chemaly told me that online harassment may sometimes be more intense and may involve many more people than in-person harassment, but "its origins, its roots, its foundations don't come from the technology. They come from the culture."

• • •

Brooke is almost always the only woman on a jobsite. Women make up 5.8 percent of welding, soldering, and brazing workers in the United States. For Brooke, misogyny is not an abstract concept that requires a philosophical definition. It is flagrant. She told me she wishes there were more women in the trades, and she'd like to be a trailblazer for them. Then she said she was worried that sounded narcissistic.

I told her I didn't think so.

I asked if maybe some of her choices have been attempts to prove to others what she can do, to prove people wrong. If they tell her she can't do something, then—

"I have to do it," she jumped in.

In 2019, Brooke decided to join a union. She was tired of navigating the job alone. The union had clear rules about what she was expected to do on a jobsite and what was not allowed. With a union, she would have health insurance and a pension. As much as she loved welding, she worried about the health hazards. The welding arc produces damaging radiation, and Brooke could feel the stress on her eyes. She wanted to explore a trade with fewer health risks. She decided to apply to two unions, one for big rigs and one for plumbing.

She was in line to submit her plumbing paperwork when she heard it: "Look at this freaking chick. She wants to become a plumber," she remembers the men snickering. In the United States, 2.2 percent of plumbers, pipefitters, and steamfitters are women.

In her head, Brooke thought, "Yeah, I'm going to be a master plumber!"

Brooke became one of a handful of women in a union with approximately a thousand members. She became a plumber apprentice.

Nearly two years later Brooke attended the Tradeswomen Build Nations conference in Las Vegas. She told me it was a diverse gathering. She met women of different trades but also different races, ethnicities, and sexual identities.

"This conference could not handle the amount of things that were thrown at them," Brooke said. "I remember . . . a woman was saying that they told her to jump off a building. These guys were like, 'You're pathetic. Just jump off. You're a carpenter. You think you're going to make it?' The things that I heard made what I went through pale in comparison. And these women who were answering the questions were like 'Listen, sister, we know. We hear you.'"

Afterward Brooke was angry. She had an Instagram account that she used sporadically to share pictures of her life, but she decided to transform it into a page where she could talk about her experiences of being a woman at work.

The abusive comments promptly poured in. She used hashtags to draw an audience, attracting people from all over the Internet. Men would harass her, telling her she wasn't a real plumber, and they didn't want to see a woman pretend to plumb. Women would harass her, telling her they could see her makeup and knew she was working in the trades only to attract male attention. It was the worst when Brooke tried public vulnerability, posting about a day so difficult she had cried in a porta-potty.

"I got a lot of flak for that. I got a lot of like 'Oh, you should just jump on a pole.' Always lesbian comments. Things like that. Always. Which is what it is. I'd rock it. I wouldn't have any problem with that. But that's not who I am."

When Brooke began posting online, the harassment was no longer bounded by the worksite. Now it was diffuse, in some ways more cruel, and there was no boss to intervene, no consequence other than a block, which wasn't much of a consequence at all. It was everything she already knew, and it was more, because there were more of them. It's the volume of hate that gets to her, she said, the sheer number of people who come into her comments to tell her she is a joke. It is difficult, she said, to dismiss a laughing chorus.

Chemaly told me that while online abuse is an extension of preexisting cultural norms, it is not a "linear extension." Rather, "it's an exponential extension because the tactics, and the behaviors, and the abuse, and the violence, and the threats, all of them exist already, but when you move them into this technology environment, their quality changes, and so the experiences are more profuse." She continued, "They are longer lasting. They are more intense. They often involve many, many more people, so it becomes something new and different in and of itself."

In a special issue on online misogyny published by the journal *Feminist Media Studies,* authors Debbie Ging and Eugenia Siapera wrote that "digital technologies do not merely facilitate or aggregate existing forms of misogyny, but also create new ones that are inextricably connected with the technological affordances of new media, the algorithmic politics of certain platforms, the workplace cultures that produce these technologies, and the individuals and communities that use them."

Brooke's experiences with misogyny online had a cumulative effect. She told me that a year before we spoke, she was anxious and depressed. She had thyroid issues. She felt worthless. She was struggling in her relationships with the men around her. She resented her husband, also in the trades, who didn't have to deal with the abuse. She resented the men in her union, who didn't have to deal with the abuse. She resented the well-meaning suggestions people made: Stop wearing makeup. Grow thicker skin. Ignore it.

You can't ignore it.

"You absorb it," she said.

Veering from suffering's well-worn script, she began to do things differently, offering herself a ferocious tenderness, occasional respite, authentic connection, presence. Through small acts, she began to engineer a way of meeting her own needs. She attended to the dimensions of body and self that fell victim to relentless denigration and ridicule. She finally decided to give herself the care the culture was intent on denying.

Self-Care

For our conversation, Brooke and I met at a small health food store, her suggestion. The owner wrote on the store's website that she wanted to empower her community with more "health and wellness options." The About Us section reads "The path to changing the world begins with changing ourselves."

It can be difficult to tend to yourself in a culture that demands you hate yourself, that profits from the labor you do to constantly reshape yourself, and that trains you to tune your body, your temperament, your very being, to the desires of men.

Such dispiriting demands can turn us toward unhealthy habits: substances that take the edge off, food that does not nourish, even greater immersion in the technologies that are profiting from our pain.

In her book *How Emotions Are Made*, neuroscientist Lisa Feldman Barrett writes that one of the best and most basic things we can do for our health and well-being is to keep our body budgets in good shape. It is all the obvious things—getting enough sleep, eating healthfully, moving, connecting, being touched when we want it and invite it. Barrett writes that crying can help, too, when it slows breathing, which can promote calm.

There is no shortage of self-care guides to help women manage their online abuse. PEN America's field manual on navigating online harassment, which has a section on practicing self-care, recommends strategies like bodywork that involve physical check-ins and body-centered therapies like massage, as well as listening to music, maintaining religious or spiritual practices (if you have them), and spending time with people you love.

Guides can be useful, but I've always found their tidy recommendations too prescriptive. Defining a self-care practice is personal and always subject to change. Not everyone enjoys yoga; I don't. For some trauma survivors, sitting still in meditation can feel impossible. For some of us, caring for our skin or our hair is a necessary part of loving our bodies. Others will find these practices frivolous or fraught. Broad conceptions of self-care cannot capture our unique relationships to ourselves.

Critical justice scholar Loretta Pyles wrote her book *Healing Justice: Holistic Self-Care for Change Makers* after experiencing burnout from years of working with survivors of domestic violence and pursuing a career as an academic. Developing a self-care practice, she told me, was necessary for her personal well-being and her work, and honing it took trial and error.

"We want something that's going to help us be more present in our bodies. It's going to help us be more present with our environment, with a greater sense of meaning or being, something a little bit greater than just our small self," she told me.

Pyles said that ideally self-care helps develop a sense of steadiness, of pleasure, of joy. It should not be something so strenuous and time-consuming that it becomes another thing to beat yourself up about.

"That's so much a part of our capitalist culture, that it's got to be something hard that wears us out or challenges us or pushes us to our limits," she told me. "What if it's just something fun and joyful that we did as a kid that we want to pick back up? Maybe it's making a collage, just cutting pieces of a magazine out and sticking them on a piece of paper in expressive ways to find your center."

The women I interviewed for this book had diverse approaches to self-care. Some returned to old hobbies and others cultivated new ones. Many people spoke with me while on a daily walk. Women told me they disciplined themselves to get off their phones and computers. Victims' rights lawyer Carrie Goldberg said that in her office, she and staffers intentionally created a space soaked in light, with comfortable sofas and yoga mats for meditation.

Many women also journaled, a way to get the pain out of the body and onto the page. Sometimes I considered this book like a public journal. Writing to you has been therapeutic for me.

Writing therapy was pioneered by the social psychologist James W. Pennebaker. In the 1980s he conducted a survey of about eight hundred students and asked the participants if they had had a traumatic sexual experience prior to the age of seventeen. About 15 percent of the students said yes, and those were the same students who had much higher rates of health problems. Follow-up research suggested the issue "was not having a sexual trauma per se." Those most likely to suffer were keeping their trauma a secret.

Pennebaker wanted to know what it was about secrets that was so toxic. His theory was that "holding back powerful emotions, thoughts, and behaviors . . . was itself stressful," and he was curious as to whether expressive writing could help.

He ran his first experiment in 1983 and found that students assigned to write about any type of trauma for fifteen minutes a day for four days ended up in better health. Over the next six months, they visited the student health center half as much as the control group. By the mid-1990s, a consistent literature had developed on the connection between expressive writing and health. In 1997 Pennebaker wrote an article in *Psychological Science* summarizing the research on expressive writing to help researchers and clinicians understand what parameters made the exercise most successful.

More than fifteen hundred expressive writing studies have now been conducted in laboratories all over the world. Pennebaker would later write that he believed his paper was impactful because it touched on questions fundamental to the human experience, on why it is so valuable to explore and understand the issues that weigh upon us.

Sometimes we do not always have the language to express our pain. When words escape us, movement can help us emote what is unsayable or unsaid. Alexandria Onuoha uses her body to communicate legacies of anguish, as well as cultural pride. She has danced her whole life, and now she teaches young girls to dance. Once I stumbled on a TikTok video of a friend who had survived an abusive marriage and was attempting to survive a contentious divorce. She danced on tiptoes, kicking her legs, spinning, running her hands down her chest, moving to encourage release.

Self-care is a tenderness for ourselves when we come up against a hard world. When we are exhausted, it can be difficult to create time for it. Sometimes we must carve it into small spaces. As Audre Lorde wrote in the essay "A Burst of Light," she would check in with herself during the briefest periods, such as "sitting on the Staten Island Ferry on my way home, surrounded by snapping gum and dirty rubber boots."

Brooke said she is practicing self-care through consistency. She is working out, nothing big, twenty minutes on the bike, "half-ass form," she said. She is paying attention to how she eats. She journals her gratitude. Little things, like being grateful she didn't stub her toe on that piece of wood sticking out of the wall. Grateful she has the energy to exercise. Grateful she slept well that day, that she got the laundry folded.

She talks to herself too, tells herself the things she needs to hear so that she can say the things aloud that she feels other women need to hear. She recharges by spending time with friends and especially her daughters.

Scholar Jessie Daniels told me the way many people think about self-care is so individualized and atomized that it distracts from what's really healing, which is connecting with one another, carrying pain together, working to eradicate violence together. When Brooke talked to me about self-care, it was clear she was not talking about the billion-dollar industry of self-care, a rich white feminized version of beauty or health. She talked about care that allows us to connect with ourselves so deeply, to love ourselves so completely, that we can care for something more than ourselves. She hopes to serve as the mentor she never had. "To get ready for this next stage in my life," she told me, "I have to be the healthiest I can be."

Listening to Brooke, I reflected on my own approach to self-care, though I realize I would never have labeled it as such. I see a therapist most weeks. I try to eat a balanced diet, and I sleep as much as young children will allow. I run and I write. Sometimes I run to write, especially when I am stuck. I love running in Central Park. I love to watch the most disciplined runners, their taut bodies and enviable workout gear. I especially love the novice runners, the ones laboring to stay with the incline, holding their phones for dear life. I love all the difference in one place. But then I remember, too, what Clint Smith wrote in *How the Word Is Passed,* that parts of Central Park were built only "because several generations ago hundreds of Black people were violently forced from their homes." When I run, when I keep myself strong, I think, What world does this body work for?

• • •

After years of working for men, in 2022 Brooke got a job with a woman-owned company. She thought, finally, things would be different. But her first day on the job, the site didn't feel different. Men made their typical comments. She met her new employer at an industry event. Brooke was nervous but curious. She introduced herself.

The woman shook her hand and asked, "What are you doing on my jobsite? You're way too pretty to be on my jobsite. I'm sure you heard how I feel about women on the jobsite. I feel like they're always looking for problems. They're always looking for attention."

One week later Brooke was laid off.

"Fuck this," she thought.

Shortly afterward Brooke opened her own LLC. She's starting a women's committee at her union. She told me she now wants to launch a YouTube channel with content on the experience of being a woman in the trades. She is determined to keep herself strong for what's ahead.

Several months after our interview, I saw Brooke at a social event. We got on the subject of bad days. She told me about an awful string of them: her minivan broke down, she ran over a possum, the guys were horrible on the job. When she struggled to lift some pipes, one of the guys said to her: "I don't know why you're trying so hard. You're a woman. No matter what you do, no one is ever going to take you seriously."

I can't remember if I yelled or snorted in response, but I knew Brooke wouldn't let it stand. I asked her, What did you say?

"I talked shit," she said. "I asked him if that's how he wanted to talk to his future boss."

CHAPTER 9

Women, Too

WELLESLEY COLLEGE IS the premier women's institution in the country, a private liberal arts school that boasts Hillary Clinton, Madeleine Albright, and Nora Ephron as alums. At Wellesley, women don't have to pretend to be anything less than passionate and impressive. The school's motto is "Non Ministrari sed Ministrare," which translates to "Not to be ministered unto, but to minister," stressing its commitment to leadership and service. The school believes in the power of its graduates to "envision the world in which they want to live and to take action to make it real."

The first time Lindsey Boylan visited Wellesley, she knew there was no other college she would attend. Born in 1984, Lindsey had paid close attention to women's political gains, seen a record number of women senators elected in the "Year of the Woman," watched as Albright became the first woman secretary of state and Clinton became a U.S. senator.

These high-achieving women seemed different from those in her own life, Lindsey told me. Her grandmother, aunt, and older sister all struggled with mental illness and eventually lost custody of their children, losses she indicated were more complex than this chapter could hope to address.

"Every story that I had about women in my life was a tragic story and was a story about pain and trauma and outside forces making it next to impossible to overcome," she said. "I was unwilling to have that be my story."

At Wellesley, Lindsey saw women who were strong and empowered and had been safely nurtured. On campus, she said, no one had to navigate male flirtation, male derision, men's entitled interruptions. The most popular women at Wellesley were the most intellectually curious and driven. She called her experience "transformative."

Wellesley shows its students what a world of powerful women can look like, and then they leave. As one alum told *The New Yorker:* "Women are first-class at Wellesley. . . . When you graduate, it's hard to go back to being second-class."

After graduating, Lindsey moved to New York and began working for an urban planner, which set her on a path that would lead to former governor Andrew Cuomo's office, where she would eventually serve as a special adviser.

Lindsey had worked hard to get into that room. But the culture, she said, was so toxic, so abusive, that she was nauseous before the start of each day. There was Cuomo's invitation to play strip poker on a government airplane, the email from another aide suggesting the governor thought she was good-looking, the times the governor would touch her back, her arms, her legs. The time they were alone in his Manhattan office and when she got up to leave, he kissed her.

In December 2020, Lindsey became the first woman to publicly accuse Cuomo of sexual harassment. It had taken years for her to come forward, she said, because she did not want to be considered a victim like so many of the women in her own family and because she had grown up watching powerful women whom she admired withstand their own mistreatment. She believed there was so much you had to tolerate just to be allowed in that room.

"Maybe it's who I was raised by, or how I was raised . . . but I always unfortunately, or fortunately, equated leading as a woman with a willingness to handle pain."

Lindsey came forward with her sexual harassment accusations against the former New York governor two years after she resigned from her position in his administration. At the time, she was running for Manhattan borough president, following an unsuccessful bid for Congress. She had already tweeted about Cuomo's toxic and abusive workplace culture, but she decided to share details of her sexual harassment after another woman reached out to her to privately disclose that she too had been harassed by the governor. Cuomo had become a popular national figure during the coronavirus pandemic. Lindsey worried that he would only become more powerful.

Lindsey was in the car with her husband and daughter when she began firing off tweets about Cuomo's sexual harassment. They exploded. Two months later she provided further details in a *Medium* post.

But it was only after her disclosure, she told me, that she experienced what she perceived to be the greatest cruelty, worse than the worst thing a man had ever done.

Online, a group of women banded together to eviscerate Lindsey. They called her a "bully," said she was "pathetic" and "cringeworthy," "evil" and "shameless," a "lower-grade version of Amber Heard." They accused her of "rotten spoiled gold-digging Karen meltdowns." They said she was "a liar."

"It was heartbreaking," she said.

Lindsey was crying when she told me that. She said she was glad that she was crying, so I could hear it.

"What transformed my relationship to my life, and the people in it, and everything in my world, was this experience of being dragged and harassed and objectified because I came forward. Not because I was abused," she said. "Some people hate women. And some of those people are women."

Women Abusing Women

When I first conceived of this book, I assumed that abuse perpetrated by men was the only problem worth addressing. Most of the attacks I experienced online appeared to be perpetrated by men, which I gleaned from the names they attached to their tweets and emails, but also from their sexualized language and other crumbs about gender they left in their messages, despite their encrypted addresses and other efforts at concealment.

Mainstream conversations around women's online abuse tend to focus on trolls, white men, and misogyny. But the more abuse I experienced, the more I saw where women were guilty, too. When I covered the Johnny Depp and Amber Heard trial, the most disgruntled and abusive emails came from other women, many of whom claimed to be survivors of domestic violence themselves, who insisted it was obvious Heard was lying. Many of the women I interviewed or talked to informally for this book said women were among their most vicious attackers. One analysis found that on Twitter, women are almost as likely as men to use the terms *slut* and *whore*, both casually and offensively.

You could spend a whole book unpacking the complex phenomenon of women perpetrating harm, especially against other women. All women—conservatives and progressives, privileged women and marginalized women, avowed feminists and those who would never refer to themselves as such—who operate in patriarchal societies are susceptible to internalizing misogyny and tormenting one another to claim the scraps that an unjust system of power provides.

One Wellesley student who recently graduated told me that while the college is a place where women can flex their intellect and ambition, the academic culture also led to a competitiveness between women that she found "toxic."

In 1952, Group Analytic Society International was formed to study and promote group analysis, a method of psychotherapy that originated with one of its founders, S. H. Foulkes, who used it to treat soldiers after World War II. Group analysis is interested in the relationship between the individual and the groups of which they are a part, as well as the wider social context.

In 2021 London-based psychotherapist and group analyst Sue Einhorn was asked by the Group Analytic Society to give the annual Foulkes Lecture. She told me it is a bit like receiving an Oscar.

Einhorn put forward the topic of her lecture, which would explore how internalized misogyny affects relationships between women. But some organizers were unhappy, she said. She received emails telling her that the topic was not psychoanalytic enough, that she had become too political. But Einhorn told them her topic wasn't up for debate.

Einhorn prevailed and would later tell her audience how women uphold patriarchy by policing one another. This hostility is rooted in childhood, she told me, when women are taught to conform to avoid the violence of men. When a woman does not conform, other women may perceive her as a threat, or they may lash out at her because they carry their own shame about not having spoken out against violence committed against them.

Women fundamentally fear men, Einhorn said, fear being killed by men. They may not fear the individual man in a household, but women are always aware of the threat of male violence. This, she said, is why mothers teach their daughters how to dress and behave, how not to arouse the anger of men or the unwanted sexual attention of men. When I reflect on my own upbringing, I believe nearly every lesson on propriety was essentially a lesson about safety. Einhorn has studied the playground behavior of young children in the U.K. and has observed girls trying to work out very early who the most powerful people are, how to fit in and align with them.

After Lindsey came forward, other women did, too. The office of the attorney general investigated the accounts of eleven women, concluding that "the Governor engaged in conduct constituting sexual harassment under federal and New York State law." In August 2021, Cuomo announced his resignation. *The New York Times* reported that of the more than 230 people who donated to Cuomo's campaign after his resignation announcement, three out of four appeared to be women.

"Online, they have banded together in Facebook groups and on Twitter in a tireless campaign to boost his legal team and muddy his accusers," the *Times* reported. "Some hold regular Zoom meetings; others sell Cuomo-related merchandise (T-shirts emblazoned with the word 'allegedly,' among other items). . . . Twitter data and interviews with some highly engaged Cuomo supporters suggest that the number of active participants is probably somewhere in the hundreds, though a far smaller set of accounts produces much of the content."

In 2023 the *Times* reported that the grassroots activists smearing Lindsey and other accusers were quietly led by another woman: Madeline Cuomo, Andrew Cuomo's sister.

The AG's report on Cuomo also uncovered how women in his office were complicit in facilitating the harassment and at times worked to discredit Lindsey. It was a woman staffer who emailed Lindsey to tell her the governor thought she was attractive. After Lindsey tweeted that Cuomo had harassed her, senior staff members in the executive chamber, which included women, characterized her as "crazy."

One member of Cuomo's former communications staff now works at Facebook's parent company Meta as a director of public affairs. She is one of the aides who suggested Lindsey Boylan was a liar.

White Feminism

Lindsey told me she felt betrayed not only by women but by feminism. Cuomo's abuse and the aftermath forced her to reckon with her own feminist values and the women she had long admired, some of whom she had looked up to since she was a girl.

Lindsey sees at least some of what she experienced as contiguous with *white feminism,* a sometimes nebulous concept that is often used to explain feminism that is exclusionary toward Black women, women of color, and women with other marginalized identities. In online discourse, some have used the term more expansively to indict a brand of feminism that does not confront structural systems of oppression but rather works within them, achieving power for only some women, namely white ones.

After #MeToo, Koa Beck, a journalist and former editor in chief of *Jezebel,* was paying close attention to conversations about white feminism online and noticed that no one was operating from a single definition. When she decided to write a book about white feminism, she began with her own loose description, which she refined through exhaustive reporting on feminist movements throughout history. In *White Feminism,* Beck defines it as "an ideology" that envisions gender equality as something achieved through "personalized autonomy, individual wealth, perpetual self-optimization, and supremacy. It's a practice and a way of seeing gender equality that has its own ideals and principles, much like racism or heterosexism or patriarchy. And it always has." She calls it a "state of mind" and argues that it can be practiced by anyone, "replicating patterns of white supremacy, capitalistic greed, corporate ascension, inhumane labor practices, and exploitation."

Former Meta COO Sheryl Sandberg has been said to practice white feminism. Sandberg is deeply admired by many women in technology, and her book *Lean In* became a bestseller. But she has also been criticized by feminists who argue that her book minimized the structural issues that keep women from succeeding at work. Others have pointed out that she spoke about women's empowerment while the product she helped run harmed women. Lindsey brought *Lean In* to the hospital with her when she gave birth to her daughter. She said back then, in those tender early days of motherhood, she believed what Sandberg had written, that if she could remain aspirational and overcome her insecurities, she could gain real power.

Lindsey had long admired Hillary Clinton, who has also been called a white feminist. Clinton came closer to the presidency than any woman before, but in 2016 she suffered a stunning loss to Donald Trump, a candidate who in 2005 was recorded saying that when he wanted women, he could just "grab 'em by the pussy."

Lindsey had fundraised for Clinton's presidential campaign and for an election night party at Wellesley anticipating her win, but after the allegations against Cuomo were made public, while Clinton issued a statement saying they were "difficult to read" and "raise serious questions," she did not call for Cuomo to resign. Lindsey had been warned about Cuomo's behavior toward women when she took her job in his administration. She found it hard to believe that Clinton, who in 2018 backed Cuomo for a third term as governor when he was challenged by actor Cynthia Nixon, did not know who she was endorsing.

"Is it the fault of really powerful women that I've admired my whole life that sexual harassment and abuse of power continues to happen? No. But they're certainly part of the picture," Lindsey said.

Koa Beck said it more pointedly. "White feminism is about personal gain at all costs," she told me, "even if that cost is other women."

After Lindsey disclosed Cuomo's harassment, his office leaked to members of the media what it claimed was her personnel file. The attorney general's investigation found that this leak constituted illegal retaliation and was part of a broader organized campaign to discredit Lindsey. In a piece for *The New Yorker*, reporter Ronan Farrow wrote that the file, which Lindsey has never seen, reportedly contained allegations that she bullied women who worked with her, some of whom were Black.

Lindsey acknowledged to Farrow that she was involved in

confrontational encounters in Cuomo's office, but she and other former staffers said those encounters took place under the umbrella of a broader toxic culture. Lindsey said in the three years she had worked for Cuomo, she had never once had a performance review. While she viewed the allegations as part of an effort to discredit her, she also would not outright refute them. She told Farrow, "I don't want to take anything away from a woman that may have had a negative interaction with me."

When Lindsey first agreed to speak with me, she told me she would not discuss the allegations, but given her indictment of white feminism, it felt like we needed to. Lindsey said she responded to Farrow the way she did because she recognized a blanket denial that she had ever perpetrated harm against another woman, particularly a woman of color, would be an evasion of responsibility.

She told me she thought about a Black woman who might have read the *New Yorker* piece. "She doesn't know me, she's not connected to me, and she reads that. And if I say, 'No, that didn't happen,' or 'This is different,' or 'I don't know,' or 'This seems unfair,' instead of, 'I want to take responsibility and I don't want to cause harm,' if I say anything else, then I could actually be causing her harm."

When abusive power structures are upheld, everyone within that structure is participating in something harmful—sometimes unwittingly, sometimes strategically. Beck wrote in *White Feminism* that some of the strongest reporting on #MeToo revealed "the layers and layers of assistants, colleagues, managers, business partners, HR departments, board members, and executives who helped sustain a workplace culture where this type of predation was enabled."

It is not always safe to challenge power. Many women who witness abusive situations, who are being abused themselves, are fearful that if they confront or expose violence, they will lose clout and credibility. They have legitimate fears around retaliation. Women who expose workplace harassment often find that their disclosures are met with further abuse. According to a report from the National Women's Law Center, more than seven in ten survivors who experienced workplace sex harassment faced some form of retaliation. Women who come forward often face shame, ridicule, and character assassination. They face it from men. They face it from women, too.

Purpose

Three months after the September 11 attacks, when she was seventeen, Lindsey organized an event at the Newseum in Washington, D.C., for a panel on Muslims and civil liberties. During her senior year at Wellesley, she flew south after Hurricane Katrina and helped gut homes. She fundraised for the displaced, curious about how to rebuild in an equitable way. She watched as white residents fared better than Black ones, and the poor fared worse than the middle class. She wondered: "How do you center people who are closest to the pain?"

Lindsey said she always felt government existed to improve people's lives. It's why she took the job in the governor's office and why she twice ran for public office herself.

Many of the women I spoke with stressed that they stayed online despite their abuse because they believed that what they were doing mattered. Many of them said remembering the value of their work helped them to reframe their abuse as something worth enduring. Alexandria Onuoha is committed to sharing her research on fascism's impact on Black girls. After harassment cost her a job, Dr. Leah Torres eventually found work at another women's clinic. Disinformation researcher Abbie Richards said that despite all the abuse, she feels "extremely privileged to have meaning in my work." When Dr. Céline Gounder emailed the chancellor of UCLA to inform him she was being harassed by a man with a UCLA email address, she wrote: "Despite the abuse, I have continued to speak the truth about vaccines, as is my professional duty."

I have also found purpose animating. As a journalist, I value a free press and believe our greatest calling is to expose injustice. But any resolve is susceptible to doubt. The persistent online abuse I experienced in my role would sometimes make me wonder whether I was accomplishing what I set out to do, whether what I set out to accomplish was worthwhile in the first place. Detractors can be quite insistent in their efforts to diminish you, and women are practiced in diminishing themselves.

Neuroscientist Stacey Schaefer, who studies purpose in the context of her work on emotional determinants of health and well-being, told me I could not address the concept of purpose without reading the Austrian psychiatrist and psychotherapist Viktor Frankl. Frankl was imprisoned in concentration camps that killed his parents, his brother, and his pregnant wife, Tilly. Following his imprisonment, he published *Man's Search for Meaning*, which explores a psychological approach he called "logotherapy." It focuses on helping people cope with emotional difficulty by cultivating a sense of purpose and meaning.

It was a theory Frankl developed prior to his imprisonment, but he tested it in the camps. He observed that men who appeared physically strong didn't necessarily survive better or longer than those of a "less hardy make-up." Something about people's "inner selves" seemed to make the difference.

"There is nothing in the world, I venture to say," Frankl wrote, "that would so effectively help one to survive even the worst conditions as the knowledge that there is a meaning in one's life. There is much wisdom in the words of Nietzsche: 'He who has a *why* to live for can bear almost any *how*.'" Part of purpose and meaning, he believed, involves setting some worthwhile future goal. One of his goals in the camps was to survive so he could lecture about his psychological findings.

Schaefer told me people who have higher levels of purpose in life show a faster recovery from exposure to unpleasant information, suggesting it is "a key protective factor." Studies show that purpose is a component of psychological well-being and may add years to a person's life. Schaefer's research has found that having purpose can help a person recover more effectively from stress and trauma.

Lindsey's sense of purpose shows up in many places in her story. She grew up knowing she wanted to use her intellect for good, and her career has been guided by the belief that government can make the world more equitable. She had clear purpose when she decided to disclose her harassment: to prevent Cuomo from abusing more women.

Purpose kept Lindsey motivated, despite her abuse, but sometimes the desire to accomplish her goals led her to minimize harms perpetrated against her. Other times she refused to acknowledge the harm at all.

"My dream for my whole life, as early as I can remember, was to get into politics so I could help change the world in a better way for women," she said. "I was powerful. And then I had this incredibly abusive person and incredibly abusive system around that person. I was unwilling to let it take anymore from me."

Lindsey told me the last few years have been exhausting. The abuse, the retaliation, the harassment, seeing herself on the cover of every tabloid and newspaper, being afraid to walk outside, feeling despair. She is mostly resting now, she said, though she has plans for the future. She is writing and has not ruled out another run for elected office.

During Frankl's imprisonment, he concluded that a person could find meaning in work but also in love. Two men in the camp were suicidal, he wrote, but he was able to dissuade them from acting. They told him "they had nothing more to expect from life. In both cases it was a question of getting them to realize that life was still expecting something from them." One was a scientist who had written a series of books that needed to be completed. The other had a child who was waiting for him.

At the height of her harassment, Lindsey tried to shield her daughter, then in the third grade, from her stress. The girl understood that her mother was visible and in some ways was being maligned. But Lindsey tried to dismiss it. She'd tell her, "Don't worry about that. That doesn't matter." Lindsey wanted to appear strong. Maybe even impenetrable.

With time, she reflected on how the past was repeating itself. When she was young, Lindsey had thought that to become a woman of power, she would need to deny her own pain. But she did not want to pass on the lesson, so she shifted her approach. Now she tells her daughter, it does hurt, it is hard.

On one particularly difficult day, she asked her daughter, "Do you know why I do it?"

By now there had been enough conversations that the girl knew.

"Yes," she said. "Because of me."

CHAPTER 10

A Place All May Enter

IN ONE OF her earliest videos on TikTok, Marissa Indoe is on a hike. She is filming her surroundings, so we don't see her, only her exploring feet. She steps over glossy wet rocks and fallen leaves, burnt orange and daffodil yellow, her body moving between stones, between the life tucked between them. She pauses near a waterfall, then climbs high to reveal a blanket of green forest and cool blue sea. She walks beneath a canopy of trees to where two branches stretch not up but across, as if reaching for others in the grove. She crouches at the edge of the water, where small waves tease the shore.

Marissa's mother is Anishinaabe. Her father was white. She is light-skinned and blue-eyed, with long chestnut hair that she often wears in two braids. The day of the video Marissa was hiking in northern Ontario around Lake Superior. Marissa told me she started her TikTok account to stem a suffering, to heal from traumas—those made in her lifetime and others buried deep in her genes. Marissa's platform gestures toward beauty and tenaciously resists erasure.

She has filmed herself creating and displaying Indigenous art—a painted field of crimson flowers, earrings of careful beadwork and braided sweetgrass. She has posted videos on cultural appropriation, on how status cards work and when they can be harmful. She has spoken directly to the camera about how being an ally to Indigenous people, Black people, Asian people—any people who are discriminated against—requires more than participating in the latest TikTok trend.

Marissa told me the minute she started posting, she was inundated with hate. She suspects her platform grew on that currency.

She said some people hated her because she was Native, and other people hated her because she was not Native enough. Many of her videos stress that mixed-raced and white-presenting Native people are Indigenous, too. People were especially punishing when Marissa talked about Indigenous history, and particularly about the harms of residential schools.

In 2021 evidence suggested the remains of two hundred children were found on the grounds of a former residential school in Canada. Marissa read stories about the Kamloops Indian Residential School with a knot in her throat. For more than a century, Indigenous children had been forced to attend Canada's residential schools, most of which were operated by churches. At the door, children were forced to leave behind families, histories, culture. Many were beaten and sexually abused. Some had their tongues pricked with needles when they dared speak their Native languages. Many never returned home.

After reading the headlines, Marissa saw a video of a priest talking about the graves, defending the schools, claiming that within them "good" had still been done. She took the video to TikTok and excoriated it.

Hateful comments poured in. Residential schools were necessary, people told her. She should be thankful, they said. Without these schools, Indigenous children would not have learned English. Without them, she would be savage.

Marissa cried to her mother. She cried to her therapist.

"We had just found two hundred graves of children who were murdered at those schools. And then people were coming and telling me, 'You need to be thankful.'"

These were practices that had tried to rid her people of everything they are. She could find no gratitude for that.

White Western Values

It is said that the Internet is a mirror. It does not create a new reality so much as it reflects reality. It carries a history. When we look at the way the Internet was formed, how it runs, and who it harms, we find a familiar American story.

Online, Indigenous people use social media to amplify their voices and form powerful collectives. They used #NoDAPL during the protests against the Dakota Access Pipeline. (Most journalists did not report on the in-person protests until it gained traction on social media.) They used #MMIW to raise awareness about the crisis of missing and murdered Indigenous women and two-spirit/queer people. They used #idlenomore to protest the Canadian government's dismantling of environmental protection laws, and this remains an ongoing transnational Indigenous movement. But as Steve Elers, Phoebe Elers, and Mohan Dutta write in *Indigenous Peoples Rise Up,* there is also "a divergence between Indigenous worldviews and the neoliberal ideologies underpinning the design of social media." Users often engage in free labor, driven by individualistic ideas of competition (more friends, more followers), and the companies who run these platforms are not transparent about how collected data is used, especially with regard to surveillance, or about how algorithms are built to manipulate behavior.

In her book *As We Have Always Done: Indigenous Freedom Through Radical Resistance,* Michi Saagiig Nishnaabeg scholar Leanne Betasamosake Simpson wonders whether "the simulated worlds of the Internet are simulations that serve to only amplify capitalism, misogyny, transphobia, anti-queerness, and white supremacy."

Online abuse takes place on platforms largely designed by white men, built with white logic, and imbued with white Western values—individualism, capitalism, free speech. Social media has given Marissa and other BIPOC influencers a platform to engage in Indigenous empowerment, but it has also become another site of violence. Many Indigenous users and activists online are grappling with what it means to engage in decolonization work on systems guided by values that stand in contrast to many of their own.

When white men conceived of the early Internet, cyberspace was regarded as something they could claim and explore, and they warned against any interference in that goal. Look at John Perry Barlow's 1996 "A Declaration of the Independence of Cyberspace."

The first time I read his paper, I was captivated by his argument, both hopeful and arrogant, that users were creating "a world that all may enter without privilege or prejudice . . . a world where anyone, anywhere may express his or her beliefs, no matter how singular, without fear of being coerced into silence or conformity."

The second time I read it, months later, I was struck by language I had not noticed before, the references to "nature," "natives," and "frontiers."

Barlow rejected laws that claim to regulate speech: "These increasingly hostile and colonial measures place us in the same position as those previous lovers of freedom and self-determination who had to reject the authorities of distant, uninformed powers." Speaking to territorial governments, he said, "We must declare our virtual selves immune to your sovereignty."

But if the Internet is conceived of as an entitlement, as a frontier of perpetual opportunity, who must be harmed, eliminated, uprooted, and erased in order for that vision to unfold?

In scholar Whitney Phillips's book *This Is Why We Can't Have Nice Things,* she writes that when Barlow declared the Internet a place free from the tyranny of governance, it functioned as more than a Declaration of Independence: it was "Manifest Destiny version 2.0. To these early adopters—the vast majority of whom were white males—the Internet was . . . something to harness and explore, something to *claim*."

Phillips argues that trolls "echo Barlow's utopian vision." She conducted her research on a specific subcultural variety of trolls born of the early to mid-2000s on 4chan. While she did not study the trolls active today, her findings could be applied to contemporary antagonists.

Phillips does not believe trolls are some spontaneous, opaque phenomenon; rather, they reflect an entitlement "spurred by expansionist and colonialist ideologies," an entitlement to claim the Internet, to do with it as they please. Trolls want to win, she writes, to dominate, to push boundaries. Their method is adversarial, and their behavior privileges Western values and male-focused thinking, rationality over emotion, dominance over cooperation. Phillips argues that trolls are animated by the desire "to go further, to go faster, to go where no one (well, no one deemed important enough to count) has gone before—this is, at least is said to be, the defining feature of Western culture."

In her book *Gaming Democracy,* scholar Adrienne Massanari argues that "underlying the histories we tell about Silicon Valley is a set of very American values that shape the way that tech companies are structured" and "platforms are regulated." Often referred to as techno- or cyberlibertarian, this ideology is fundamentally based on principles that rely on "Western frontier myths as a defense against critique, political intervention, and regulation on behalf of the public." Men who defined early Internet culture, Massanari writes, warned that "restrictions on how companies 'settled' the Internet's 'uncharted lands' would suppress innovation and lead to 'economic devastation.'"

American society was built through the violence of colonization, through enslavement and dominion. It was built on freedom for some but not for all. How do we reckon with technology that has connected people across continents and cultures, a cyberspace that has grown to cover the earth itself but that is governed largely by the self-interest of a powerful minority? How did the promise of a global conversation evolve into the omnipotence of a handful of technology companies who view the threat of regulating violence as more dire than the violence itself?

A White Supremacist Internet

While the early idealism of the Internet saw it as a place without hierarchy, a cyberspace where people could connect and share thoughts freely, it could not separate itself from the biases, prejudices, values, and bodies that built the technology.

It is worth noting here that while most stories we hear about the birth of the Internet involve men, tech historian Claire L. Evans writes in her book *Broad Band* that women were at the start of every important technological wave. Ada Lovelace wrote mathematical proofs that many scholars describe as the first computer programs. Women were working on computational projects for the military during World War II. The women of the ENIAC 6 programmed the first electronic computer. The feminists and organizers of the Resource One computer center created the Social Services Referral Directory, which connected social workers and families in need. And these women, Evans said, had something in common: "They all care deeply about the user," she wrote. "They are never so seduced by the box that they forget why it's there: to enrich human life."

But persistent stereotypes about what makes a good computer programmer—male, socially awkward, ruggedly individualistic—have brought us the Internet we know today, which is "the vision of a particular set of white male programmers who made something in their image," anthropologist Sareeta Amrute told me.

In 1995, 42 percent of Americans had never heard of the Internet. In 1996, Pew reported that men made up a disproportionate share of online users. Today Internet use is near ubiquitous, and now we are operating on platforms that scholar Alice Marwick said are shaped by "a white supremacist tech industry."

Around the same time Mark Zuckerberg started Facebook, he created a Hot or Not–inspired "prank website" to rank the attractiveness of people in his Harvard dorm. Part of the early conception of YouTube originated with one of its founders wanting a solution for how to watch Janet Jackson's humiliating Super Bowl wardrobe malfunction on demand. On sites like Reddit and 4chan, known as hotbeds for trolls, it was the most outrageous and dangerously offensive content that got the most attention, despite the fact that many men building these early platforms believed that the best content on the Internet would rise to the top, that users themselves could root out bad behavior, and that we could all tolerate the assumed universal risk.

Former Twitter CEO Jack Dorsey sometimes indicated he understood the seriousness of harassment on his platform, although activists said he was too slow to act. But when Elon Musk bought the company in 2022, he emphasized his own brand of so-called free speech absolutism, firing employees who criticized him and capriciously changing course on content moderation when it threatened the company's bottom line. He gutted the company's staff, including 30 percent of its Trust and Safety team, and reinstated many big accounts that had previously been banned for hateful rhetoric, including those of white nationalists and neo-Nazis.

Social media platforms may claim neutrality, but they are run by people who privilege certain users and who have failed to mitigate abuse when they believed it might upset certain political groups. A *Wall Street Journal* investigation in 2021 found that while Facebook purports that all users are able to speak equally freely, its system exempts "high-profile users from some or all of its rules." Moreover many of these users "abuse the privilege, posting material including harassment and incitement to violence that would typically lead to sanctions." In 2021 *The Washington Post* reported that while Facebook researchers pushed for aggressive changes to its software system that would remove hateful posts before users could see them, Facebook leadership nixed the plan. The *Post* uncovered a document that showed insiders had concerns about potential backlash from "conservative partners."

The people who make decisions about the content we see, who create the formulas that decide how content is delivered, hold their own biases and beliefs. As Safiya Noble writes in *Algorithms of Oppression,* algorithms are built and technology companies are run by people who "hold all types of values, many of which openly promote racism, sexism, and false notions of meritocracy."

These values drive decision-making at every level, leading to abuse as well as to censorship. A 2023 Human Rights Watch report found that after armed conflict escalated between Israel and Hamas following the violence on October 7, 2023, Meta, formerly Facebook, was systemically silencing voices that expressed support for Palestine and Palestinian human rights. Since that report, Israel has been accused in the International Court of Justice of "genocidal acts." In July 2024, the ICJ ruled that Israel's occupation of the West Bank and East Jerusalem violated international law.

Barlow may have envisioned an Internet free from tyranny, an Internet where we could all participate as equals, but in the decades since he wrote his declaration, the Internet has been organized to benefit the most powerful among us.

Marissa is an Indigenous woman who has used social media to expose the harms of colonization, but the values embedded in her content are at odds with at least some of the values embedded in the platforms themselves. All Indigenous activism translates into profit for Big Tech, "corporations controlled by white men with a vested [interest] in settler colonialism," scholar Leanne Betasamosake Simpson writes.

I spoke about this incongruity with Taima Moeke-Pickering, a Canadian–New Zealand academic and a Maori of the Ngati Pukeko and Tuhoe tribes who has studied Indigenous activism online. We talked about a radical reimagining, not of the technology itself—Moeke-Pickering believes social media platforms are vital sites of activism work—but of the structures of power that control and maintain it. When she thinks about decolonizing the Internet, she said the path is simple: keep the technology, but free it from the influence of the powerful elite.

"How would I rebuild it?" she said. "I would make sure imperialists didn't own it."

• • •

Scholar Sara Ahmed writes in *Living a Feminist Life* that "the violence that we have to survive is not only gender-based violence, or violence that might take place at home; although it includes these forms of violence. It is the violence of enslavement, of colonization, of empire. It is the requirement to give up kin, culture, memory, language, land."

Indigenous people experience some of the highest rates of violence, mental health issues, trauma, and poverty. Indigenous women face an epidemic of violence, overrepresented as victims of sexual violence, physical violence, and murder. A report from the Royal Canadian Mounted Police found that between 1980 and 2012 more than one thousand women and girls identified as Indigenous were murdered—a rate 4.5 times higher than that of all other women in Canada. The Canadian government reports that many Indigenous women, girls, two-spirit, lesbian, gay, bisexual, transgender, queer, questioning, intersex, and asexual people experience higher rates of gender-based violence, which it connects to legacies of colonialism and historical trauma.

Eighty-four percent of American Indian and Alaska Native

women have experienced physical, sexual, or psychological violence in their lifetime, according to the National Institute of Justice. They experience violence 1.2 times more than their non-Indigenous women counterparts, and research shows that 96 percent of sexual violence against American Indian and Alaska Native women is perpetrated by non-Native assailants.

Marissa grew up with stories of violence. She witnesses and embodies the legacy of colonization, grieving her grandmother's refusal to pass on the language. She has experienced intimate-partner violence. A former boyfriend once covered her mouth while she was crying so that she couldn't breathe. Another time he threw her into a door. After she broke up with him, he found her online. He emailed her, DM'd her, and harassed her on Facebook, on Snapchat, anywhere he could find her.

When Marissa first joined TikTok, she was newly sober and looking for a way to stay accountable. She had been abusing alcohol to manage triggers related to the abusive relationship she had ended. She posted sobriety updates and grew close to other Indigenous people who were also sober. The connections were invaluable, but the online harassment was suffocating. People online told her it was just in her DNA. "Drunk Indian," they called her. (In a 1996 article in the *Arizona State Law Journal,* authors Robert J. Miller and Maril Hazlett wrote that the stereotype of the "drunk Indian" suggests Native Americans are predestined to alcohol abuse, though studies argue that "problems arise from a common experience of Indigenous people being crushed under conquest and the domination of other cultures, and not from a biological predisposition to alcohol.")

Sometimes during the harassment Marissa would lie on the floor and cry. All she wanted was a drink. She said one time she had a bottle of alcohol in her Uber Eats cart, one hateful comment away from checkout.

Marissa decided it wasn't safe to center sobriety on her TikTok page, so she shifted most of her content to focus on racism and colonialism. She stopped using trending audio and started using her own voice to speak about the devastating legacy of residential schools. She embraced the tradition of many other Indigenous women and in particular Indigenous feminists, who are using social media to shift narratives and reclaim power that was stolen from them through colonial violence.

When Marissa changed the focus of her content, hate comments proliferated. They were less personal, so in some ways they felt less dangerous, but sometimes they still made her cry. White users spewed hate, told her she was wrong about her own history. Other Indigenous users enacted what Marissa called "lateral violence," especially when she would advocate for mixed and white-presenting Indigenous people.

When Marissa is abused online, she said, sometimes she thinks her body is reminded of her ex-boyfriend's rage. It's not always a conscious connection, but at night her body remembers. When the online abuse was especially bad, she'd have nightmares about her ex.

Culture

After I first spoke with Marissa, but before I had considered what it might mean to decolonize the Internet, I was interested in getting a better understanding of Indigenous healing. I had come across a 2010 paper by Washo Native American scholar Lisa Grayshield, a former academic who has argued for the incorporation of Indigenous Ways of Knowing into psychological practice. Grayshield writes that Indigenous Ways of Knowing "is an epistemology that recognizes the interconnectedness of all things."

When I called Grayshield, she was outside, and I could hear birds. I remarked on their chatter. She told me to go outside, too.

Indigenous Ways of Knowing, Grayshield said, stands in contrast to the Western paradigm of thinking, which is based on competition and productivity. In the Western paradigm, there is anger and there is fear—fear of lack, fear of not having enough, fear of not being enough. By contrast, an Indigenous paradigm is based on sustainability. Grayshield told me we sustain ourselves by recognizing that "every single solitary one of us has value, has meaning and purpose."

Grayshield told me to recognize how our actions ripple. In her 2010 paper, she quoted Chief Seattle: "This we know, that all things are connected like the blood that unites us. We did not weave the web of life; we are merely a strand in it—whatever we do to the web, we do to ourselves."

In two years, Marissa gained sixty thousand followers on TikTok, as well as a tide of hate. She turned to Indigenous wisdom to cope.

Marissa said she considered her culture's Seven Grandfather Teachings: love, respect, honesty, bravery, humility, truth, and wisdom. Each principle has its own story. Marissa found the teachings of love, respect, and honesty among the most useful. She doesn't excuse people's bad behavior, but she believes it shows that they are lost, that they too need healing. She chose to reciprocate with patience and kindness, to offer education as an antidote.

"Even when people are hateful, rude, discriminatory, racist . . . I always tried my best to handle the situation with a sense of respect, love, and honesty. Not because I love and respect racist or hateful people, but because I love and respect myself and the Indigenous community. . . . I always tried to respond to violence with education and kindness because I know what it's like to be off the correct path."

As Indigenous scientist Robin Wall Kimmerer writes in her book *Braiding Sweetgrass,* "In order for the whole to flourish, each of us has to be strong in who we are and carry our gifts with conviction." Our gifts "are not meant for us to keep. Their life is in their movement, the inhale and the exhale of our shared breath. Our work and our joy is to pass along the gift and to trust that what we put out into the universe will always come back."

Publicly, Marissa has responded to her abuse with compassion and composure. Privately, managing her anger and disgust can require a different approach. "There's also times, too, where the comments are so stupid or so unoriginal that I just laugh," she told me. "I make fun of them. I go downstairs and I read them to my mom, and we start laughing our asses off."

To cope with the abuse, Marissa also engaged in rituals. She smudged her space to clear negative energy. When she needed protection, she burned sweetgrass, which Kimmerer, a botanist, writes is a sacred plant, and which Indigenous stories say was the first plant to grow on earth. Important to many Indigenous nations and often used in ceremonies, it has "the power to focus attention to a way of living awake in the world."

Marissa makes her living as an artist. Art helps her be present. When her hands are busy, her mind is calm and open. Research shows that even sixty seconds of mindfulness a day for six months can heal brains and bodies. Brief mindfulness exercises can enhance comfort and calm and reduce pain. Mindfulness has been shown to improve emotion regulation, and it has been linked to a reduction in stress, depression, and anxiety. Critical justice scholar Loretta Pyles, who studies holistic self-care, told me practices of mindfulness and meditation and connection to nature are "every human's birthright."

Artistic expression can communicate the complexity of human experience. Former APA president Dr. Thema Bryant writes in *Homecoming* that she has "seen the way the arts can awaken people." Traditional and Indigenous art is shown to be therapeutic, to help people cope with pain and trauma. Creative visual expression is a way to make meaning out of violence, and art is a powerful tool of decolonization.

Marissa creates acrylic and watercolor paintings, as well as digital art. During the height of her online abuse, she created a piece titled *Back to Our Roots (Orange Edition),* featuring three generations of Indigenous people with hair braided together to the roots of the earth, symbolizing the connection between mind, body, and spirit. An eagle in the sky shows connection to the creator. Marissa donates a percentage of the sales of the print to an organization that supports survivors of residential schools.

She told me the piece is about "how we are still here."

• • •

While I worked on this book, I largely disengaged from social media. When it came time to draw on some of my own experiences for this narrative, I returned to my hate mail, to hundreds of screenshots of horrific comments, tweets, and DMs. Having not read them for some time, having not read anything hateful directed toward me for some time, I read them with fresh eyes and they took on new meaning. In some ways, they felt even less bearable now. Space away had helped me recalibrate, left me incredulous about what I had absorbed on so many days. Not everyone has the luxury to step away from their online life, but those of us who do can often find catharsis and sometimes new perspective.

The first time I interviewed Marissa, she told me she was still posting on TikTok but less. She wanted people to understand her culture, her history, wanted people to sit with the harms of colonialism, to desire a different future for us all. When I reached out to her a few months after we first spoke, she had gone private on TikTok and deleted the app from her phone. In her last post, she celebrated two years of sobriety. She told me her decision to leave the app was tied to the Medicine Wheel teachings, which encourage balance through mental, emotional, physical, and spiritual health.

To combat violence, many of us are working within systems that perpetrate violence. Sometimes retreating is its own form of courage.

Marissa's withdrawal from the app is a loss for her followers and for her activism, but it is also a refusal to participate in spaces that do not align with her values. These platforms may connect us, but they also profit off us, isolate us, sever us from the natural world.

Marissa is on a break from her platform, has exited Barlow's "civilization of the Mind," and for now has something she values more than his "global conversation of bits."

Peace.

CHAPTER 11

Therapy

MORGAN SUNG DOESN'T remember life before the Internet. She was ten years old when she got on MySpace and Tumblr. She was active on Stan Twitter, where bullying is the norm. People have said horrible things to her online since she was a child. She is a woman now. She told me that most of the horrible things people say to her online don't affect her much. But she wonders how the abuse changed her budding brain, how it changed who she might have become.

When we spoke, Morgan was twenty-seven. She belongs to a generation that can tend to shrug at abuse online. "This is just how the Internet is" can be said by abusers who want to abuse without consequence. "This is just how the Internet is" can be said by victims who have no idea what consequences might look like. Morgan doesn't believe this is how the Internet should be, but this is also the only world that women her age have ever known. They don't remember life before the Internet, when you would travel places and eat foods and laugh with friends and see strange things, and the only people who would know were the five people you told, at lunch, or dinner, with no one on their phones, looking down, commenting, searching.

When she was a child, she was anonymous online. When people made rude or cruel comments, it didn't feel like it was about her. She wore a mask, and her perpetrators did, too. "When I was fourteen on

Tumblr, getting into an argument with someone else who had a supernatural Doctor Who profile pic, to me, I was like, 'I'm not a real person online, and neither is that other person, because we're just anonymous people arguing.'"

But Morgan grew up and wanted to be a journalist, and everyone told her she had to be very online, so in college she ditched her old Twitter account and created a new one with her real name and face.

People said terrible things to her, but now she was a real person, and now she was afraid. She told me she doesn't believe people were built to feel this exposed.

There have been more "dumb bitch" moments than she can count. The first time she was genuinely frightened, someone posted her name, age, and neighborhood on a message board. But it was an experience in 2020 that marked a clear before and after, during the height of the Black Lives Matter protests, when she wrote a story on people toppling Confederate statues. Morgan interviewed an archaeologist on how to topple a statue safely, after protesters across the South began removing Confederate symbols that they saw as celebrating white supremacy.

Readers threatened to find her and her family. Morgan is Asian American. They threatened to send her back to her country, even though she was born in New York. They said her purple hair made her look more rapeable. They called her Antifa. The right-wing political commentator Dinesh D'Souza, who has millions of followers on Twitter, retweeted a video of Morgan discussing her reporting, called her a "smug little fascist," and said he wished she would be indicted and prosecuted "to teach her a lesson."

In response, Morgan printed out D'Souza's tweet, put it on her refrigerator, and took a smiling selfie with it that she snark-tweeted back. The printout is fixed to the fridge with a magnet of Morgan from eighth-grade picture day. "It was weird because I truly felt nothing at all," she told me. "It was like something in my brain just shut off."

Morgan doesn't remember much about the days after. She told herself it didn't matter. She told herself it was just the Internet and none of the threats were real. In retrospect, she said there were probably credible threats, but she tried to ignore them. When she talked about it with friends, she would tell it like a funny story. This is just what it's like being a woman on the Internet, she explained to people.

Six months later, she was separated from her family during the pandemic, and she remembers telling her therapist she felt unbearably lonely, the loneliest she had felt since the summer, right after the attack. And only then did she finally begin to talk, not just about what had happened but about how it felt.

"My therapist was like 'Wait, what? Why didn't you bring this up?' And I was like 'No, it's fine. I got through it. It's whatever.' . . . I had shut it off and numbed myself to it. . . . I remember the minute I started talking about it with my therapist and actually openly talking about it, not using humor to deflect, I just remember my heart rate shooting up. . . . I remember my voice started shaking and I was like 'I don't know why I'm reacting this way when I talk about it because I've been talking about it for months and it was fine.'"

Morgan wasn't fine. Her body showed her.

Therapy

When a person is suffering, concerned people around them often ask, "Have you talked to someone?" We know what this means. After Trump Jr.'s mob, I stood in my driveway, unkempt and vacant, and told a neighbor what had happened. She asked, referring to my employer, "Did they offer you a therapist?"

Many of the women I interviewed for this book are currently under the care of a mental health professional or have been at one time. Some women said therapy was an essential part of their coping. Morgan's therapist helped her learn how to more effectively process stressful situations in real time. Lindsey Boylan said it was hard to imagine how she would have handled Andrew Cuomo's harassment and the onslaught of online abuse if not for twice-weekly appointments with a trauma therapist. Sex educator Dr. Donna Oriowo sees her own individual therapist once a week and has a slew of therapist friends she connects with almost daily. Writer Lyz Lenz says that as she continues to adjust to new forms and intensities of online abuse, her therapist has helped her with "leveling up" and building new skills to cope.

In 2019, after the vice-presidential debate between Kamala Harris and Mike Pence, I wrote a story about the sexist and racist stereotypes Harris navigated. I was inundated with abusive messages, and at the time, it was the most hate I had received over a single story. The day after the debate, messages kept rolling in. I read them at my computer and on my phone. I had errands to run, and the messages trailed me. I read the story again. I examined the quality of my arguments and searched for a legitimate source of the senders' anger. These experiences can eat away at your confidence. You can begin to assume that your abusers are not the problem. You can begin to think you are.

In my first therapy session after the story ran, my therapist asked me to reframe it. I was too exhausted, so she helped. "They pile on and require you to cope with such an extreme amount of abuse that by the end of it you may begin to believe they are right," she told me. "It starts to brainwash you."

My therapist identifies as a feminist and is well versed in power and abuse, even if she has never experienced online violence herself. But many of the women I interviewed told me that when they attempted to talk to their therapists about their harassment, practitioners would struggle to understand what they were dealing with online, and even why it mattered.

Journalist Taylor Lorenz, who has been diagnosed with PTSD, said she tried two therapists who offered trite feedback on problems she felt they barely grasped. Sara Aniano, a disinformation analyst at the Anti-Defamation League, said her therapist sometimes helped but was generally too optimistic about the world. Disinformation researcher Abbie Richards said her therapist didn't understand the nuances of her experience online: the differences between trolls and credible death threats. When I interviewed Marissa Indoe, she told me she was looking for an Indigenous therapist who could better understand her lived experience.

While therapy is often touted as one of the most effective ways to cope with emotional difficulty, to better understand ourselves and reframe problematic thinking, it can be difficult to access, prohibitively expensive, and varied in quality. Some therapy can be lifesaving, some can feel futile, and some can cause harm.

White men founded the American Psychological Association in the nineteenth century. The history of mental health care in America is steeped in some of the same values that white men used to build the Internet—individualism, self-determination, meritocracy. Being well is often considered the work of the individual, and behaving well means you are a productive member of a capitalist society. Whiteness dominates mental health, which has prevented patients from accessing therapists who are best positioned to understand their lives, and which has allowed white supremacy to escape meaningful examination in therapeutic spaces. Dominant mental health care in America has a history of pathologizing women's pain rather than treating it as a symptom of a society in which women are expected to accept violence as the cost of living. Data from the American Psychological Association shows that in 2021, 81 percent of active psychologists in the United States were white, 8 percent were Hispanic, and only 5 percent were Black.

Many people are working to combat this lack of representation, including Lillian Comas-Díaz, who co-chaired the committee that developed the *APA Guidelines for Psychological Practice with Girls and Women*. She told me when a person is being oppressed, the most effective therapy seeks to understand how their experience connects to a larger system. In her view, wellness isn't compliance. It's strengthening yourself so that you may become unyielding.

Liberation Psychology

Comas-Díaz and several other psychological experts I interviewed said women can be well-served by a framework called liberation psychology, which directly addresses oppression. It is a perspective, Comas-Díaz told me, that is mostly ignored in dominant mental health, which she said is likely related to racism and xenophobia, as liberation psychology originated in the Global South. It was born in Latin America, fed by the ideas of Brazilian philosopher Paulo Freire, who wrote the seminal text *Pedagogy of the Oppressed.*

Freire believed that for people to free themselves from oppression, they had to recognize its causes, to become conscious of how oppression functioned in their own lives. Jesuit priest and community psychologist Ignacio Martín-Baró expanded Freire's idea of critical consciousness into "conscientization," one of the many methods he developed to advance liberation psychology, which focuses on the social, political, and economic factors that give an experience context.

Freire and Martín-Baró believed that oppressed people could not change their lives for the better without understanding the systems of oppression at the root of their suffering.

This consciousness also requires examining the misbeliefs oppressed people internalize. In *Sister Outsider,* just after Audre Lorde famously wrote "the master's tools will never dismantle the master's house," she references Freire, noting that he showed that "the true focus of revolutionary change is never merely the oppressive situations which we seek to escape, but that piece of the oppressor which is planted deep within each of us." If we do not grow beyond "old blueprints of expectation and response," we are susceptible to perpetrating violence ourselves through use of "the oppressors' tactics."

Comas-Díaz said the early conception of liberation psychology ignored gender. It didn't incorporate feminism or issues that impact people who are LGBTQ. In response, Comas-Díaz said, women liberation psychologists pushed for greater inclusion and birthed their own psychologies, including feminist psychology, womanist psychology, mujerista psychology, Black/African-centered psychology, and Indigenous psychology. These psychologies are distinct, but conceptually they align and challenge Western ideas of wellness.

In her book *Radical Feminist Therapy*, feminist psychotherapist Bonnie Burstow emphasized that the mental health care system has a long history of misogyny. That is why women must have a therapist who understands that "violence is absolutely integral to our experience as women." Burstow spoke out against the field of psychiatry, which she argued was deeply patriarchal. She believed women's responses to violence should not be pathologized as mental illnesses but seen as rational reactions to an unsafe world.

In a therapeutic setting, a therapist who practices liberation psychology might use the concept of *testimonio*, defined in *Liberation Psychology* as "a verbal journey of a witness who speaks to reveal the racial, classed, gendered, and nativist injustices they have suffered." They may also use the concept of *acompañamiento* (accompaniment), when a therapist accompanies a person to become more critically conscious about the relationship of their pain to social, political, and economic systems. Martín-Baró contributed to the development of the concept of *acompañamiento*, which describes the behavior of a liberation psychologist who is working with traumatized individuals, standing alongside them.

Psychologists in this tradition believe that endlessly intervening upon the individual will never bring about full healing. Dr. Thema Bryant, a self-identified womanist therapist, told me that whether a therapist identifies as a liberation psychologist or not, they are most helpful to someone experiencing oppression when they empower a patient to explore their "resistance strategy."

"We don't want to say, just cope with racism, cope with sexism, and just go light candles and take a bubble bath," she said. "Those things are good to feed yourself, but then you feed yourself for the fight, to go back and combat it, to eradicate it, to push back, to call it out, to name it."

Comas-Díaz told me that while she believes the most effective therapy helps a person become "critically conscious of what else is happening at a systemic level," that does not mean there is no merit to using other frameworks, including cognitive behavioral therapy (CBT).

CBT, widely considered one of the most effective treatments for managing a range of mental health issues, aims to teach a person skills that help them regulate their emotions. The self-help section of my local bookstore is stocked with books on CBT.

The premise of CBT is that a person's thoughts, feelings, and behaviors are related, and by operating on any one of them—but most often thoughts and behaviors—a person can reduce negative thinking patterns and change their reactions.

In a case of online abuse, CBT can teach a person to be less self-critical by helping them prepare for harassment, so they already have some scripts for what to tell themselves, including "They're coming after me because I said something true" or "They're coming after me because I am more visible than I once was" or "They're coming after me because they want to exploit me for profit or influence."

But multiple psychologists who practice CBT also told me that in the hands of the wrong therapist, CBT can be harmful. Since CBT is used to help a person think about something differently, it can inadvertently be used to suggest someone is overreacting to an experience, that their emotional pain is no more than a cognitive distortion.

"Any therapist who is not both personally familiar with and invested in making themselves familiar with the lived reality of their clients . . . can do harm with these tools by invalidating a very real experience," psychologist Emily Sachs told me.

No matter what framework a therapist uses, psychologists told me, the quality of the care is connected to the quality of the therapist-patient relationship. Violence disconnects us from other people and sometimes from ourselves. Therapy can show a patient what safe connection looks like. It can demonstrate responsible use of power. A good therapist does not believe they are all-knowing. A good therapist helps show a patient all the power and knowledge they have, too.

• • •

Morgan has become more skilled at managing her emotions, but that does not change her material reality. She still has to remain vigilant about digital safety, which creates a paranoia around what she posts and what other people post. If she's posting a picture of her foster kittens, she makes sure there are no windows in the picture, no apartment fixtures that someone could use to identify where she lives. She subscribes to services for herself and her family to wipe their personal information from the Internet.

Since the statues story, Morgan has received that same amount of harassment multiple times. She is considering what's next, and what she can manage.

Months after we spoke I tried to reach Morgan multiple times, to no avail. She came up for air once to tell me she was busy traveling. I wanted to know if she was still in therapy.

Instead, I watched her on social media, lamenting the cruel end of *Jezebel* in a post-*Roe* era. I saw her post of a picture of her latest foster kitten, wrapped in a yellow blanket, one eager claw flexed. I read an essay she published on Rat.House, a Substack about the Internet she produces with friends, on "microdosing retirement," narrating a recent break she took from writing, which led to what sounded like a fragile untangling of identity and work.

On Instagram, I saw her at Halloween, and before that in Seoul. As I scrolled, time was measured in her hair color. I saw an old post she made on Instagram of Dinesh D'Souza's tweet on her refrigerator. This, I imagine, when she was still trying to find humor in it. The caption read "early 20s fridge decor: my misspelled college diploma, photobooth strips . . . and the tweet that made dinesh d'souza mad, printed in color," followed by an emoji of a woman with one arm raised, wearing a red dress. The comments below are telling. They reflect Morgan's projected insouciance. People told her that the post was "gold" and "iconic." Some congratulated her, others called her a "legend," and one said the tweet was "like an award for excellence in journalism." Morgan replied to one commentator: "i've made it."

These replies make sense. They align with Morgan's energy, and one has to imagine that those commenting on her post felt they were giving her precisely what she wanted: encouragement, a pat on the back, recognition of her growing clout. They were responding to their friend in the moment, not to an unjust system.

Morgan's experience in therapy would shift her own thinking on the event, but those early responses from her followers reminded me of an anecdote that former *Jezebel* editor in chief Koa Beck shared with me about what she observed during her time working in newsrooms, when she and the women she worked with were regularly harassed online for stories they wrote. Women were sent rape threats, they were routinely attacked, but many seemed to wear their abuse like "a badge of honor," Beck said. They subverted the pain into pride, and she believed at least some felt that the scope of the threat was indicative of their "value on the Internet." She suspects this helped many women get through difficult experiences.

I thought this might also be true of me. Something to unpack in therapy. It's one thing to say you were abused online, it's another to say Donald Trump, Jr., was involved. I'm sure I mentioned it in most of the interviews I conducted for this book, and I'm sure when I spoke of it, some part of me felt as though it functioned like currency. *Look at the violence I have faced. Look at how much I matter.*

Don't I?

Don't we?

CHAPTER 12

Fighting Back

MIKKI KENDALL TOLD me she isn't fun to harass anymore. She said she used to be. When she was mobbed, she would tell the Internet she was afraid. When she was abused, she would tell the Internet she was angry. Now she mostly just tells people to fuck off. One night while she was asleep, a man started tweeting at her, berated her for hours, mocked her for not responding, told her she was afraid to answer. When she woke up, she quote-tweeted him and asked, Whose pile of donkey feces is this?

I can't believe you're talking to me like this, he said.

She asked if he had lost his goats in all that shit.

There is no need to be mean, he said.

We haven't even started, she told him. Give me a minute. We'll get there.

Mikki is a Black cultural critic and author of *Hood Feminism* (currently on many banned-books lists). She is a prominent voice on Black Twitter. Mikki isn't afraid to drag a troll. You might say she isn't afraid to drag anyone.

Mikki told me she isn't always nice online. She has friends on the Internet who she said are "assholes." When you're being harassed, she said, this helps. When people online come for her, she gathers a crowd. "It's a very Gen X, Black Gen X response. 'Oh, somebody is picking a fight. Let's go have a fight.'"

Mikki didn't plan to make a career online, but she was mad about a lot of things, and when she tweeted about them, it was clear that other women were mad, too. In 2013 she was mad about white feminist bloggers who she said promoted, defended, and empathized with a male writer who had been abusing Black women, including her friend Sydette Harry, for years. She tweeted the hashtag #SolidarityIsForWhiteWomen. She and the hashtag blew up.

#SolidarityIsForWhiteWomen became something much bigger than Mikki imagined or intended, exposing a historic and painful tension in the feminist struggle between white women and the Black women whose voices and concerns are routinely sidelined. In *#HashtagActivism,* authors Sarah J. Jackson, Moya Bailey, and Brooke Foucault Welles write that #SolidarityIsForWhiteWomen was "disparaged by some as divisive and unreasonable for pointing out that women of color face unique forms of oppression."

Pointing out the unique oppression that Black women face guides Mikki's speaking and writing. In 2011, the first time she was mobbed online, it was over a piece she wrote for *Salon* about her abortion. At twenty weeks pregnant, she awoke from a nap to find herself hemorrhaging. After she got to the hospital, she lay there for hours while staff essentially did nothing. A medical student showed her the ultrasound of her baby dying and asked if the pregnancy had been planned. When she was in pain, they gave her nothing. She wrote about the indignity of it, how they wouldn't even close the door so she could stop listening to the other women delivering healthy babies all around her.

After the piece ran, when she was attacked, she didn't even think to fight back. She landed squarely in a PTSD response—hypervigilant and anxious. The experience triggered her trauma, her anxiety and panic disorders. When she was a girl, her grandfather had saved her from a gunfight. When she was a woman, she saved herself from an abusive marriage.

When the mob came, Mikki didn't know what to do. Everyone had an opinion on how she should respond, she told me, and some of them felt exploitative. Some women encouraged her to talk publicly about what she was experiencing, but Mikki told me she was just trying to keep a bullet out of her kid's head. She was trying to survive. She just wanted someone who had been through it to tell her how to survive.

In the last decade, Mikki said, her approach to dealing with harassers has evolved. She went from paralysis to expressions of fear to ruthlessly dragging folks who try to build a soapbox in her mentions. She'll also respond to people online who make statements that she finds patently ridiculous.

"I had someone once tell me, 'Your children are ashamed of you. Your kids don't even know what to say about how embarrassing you are online.' And I was like, 'My kid is currently making fun of you with me right now.'"

Why the tactical shift? I asked her.

Mikki said to understand, we had to go back to childhood, to the bully who would not back down. She said she used to be a shy kid, with the nickname "Books," which she earned by being a "giant nerd." In the eighth grade, a girl started bullying her badly, and no matter what Mikki did, the girl would not relent. Mikki tried ignoring her. She tried getting along with her. One day Mikki and the girl were sitting in class together, and Mikki had had "the fuck enough." The girl asked, "What are you going to do if I hit you?" Mikki glanced at the scissors on her teacher's desk, turned to the girl, and said, "I will stab you. I'll do it as many times as I need to till you leave me alone."

The girl left her alone.

When Mikki first started experiencing harassment online, she considered it separate from anything she had dealt with before. And she continued to believe that for years, engaged in trial and error as her public profile grew, while the harassment continued—rape and death threats around the clock. But then she had an experience that changed her approach, she told me, because that harasser reminded her of her childhood bully.

Mikki was mercilessly targeted by a white woman online, and just like her bully, the woman refused to back down. No amount of ignoring, no amount of reasoning seemed to matter. The woman called Mikki's editors, she emailed Mikki's friends, she posted about Mikki on her own website. Mikki kept saying to the woman, "I didn't do anything to you," but that rationale proved futile.

"I was watching other people go through the same thing, and we're all having this spread of reactions. And then I said, 'I'm tired of being the good victim, of being the right kind of target. This isn't working for me anyway.' . . . I said, 'Okay, so it's bullying.' And it's bullying with higher stakes, but the only thing that's ever worked in my life with a bully was fighting back. So that's what I'm going to do."

Mikki said she got mean. She grew confrontational. It was only then that the woman backed down.

Later a man targeted Mikki, wouldn't leave her alone, told her he was going to "make her pay" in front of her kids, told her someone should feed her to the dogs. When the police were useless, Mikki, a military veteran, posted a video on Twitter teaching her aunt how to shoot. It was a message, she said. The man backed down.

"If I had ignored it, he would've kept going," she told me.

Mikki said the change in approach occurred around the 2016 election, and by 2019, she was regularly responding to her harassers.

"I was reminded of playing the dozens as a kid, which is this thing in the Black community where you make fun of each other. But there's a version of the dozens that goes too far," she said. "I didn't play the dozens well, because I always went too far. Well, now here was a place where it didn't matter if I went too far, they were already over the line. I could go just as far and it was fine. And even if it wasn't fine, I didn't have to care. And I'm not saying that this is for everyone, but I am saying that at some point when people are being mean to you, it's fine to be mean back."

Mikki believes having a reputation as a "mean girl" has deterred some harassers. When she makes an example of someone, it buys her months of silence.

Mikki stressed to me that she's sometimes afraid, and she has discussed her fear with her therapist, her family, her friends. But what turned a lot of the abuse down for her was an attitude of "if you knock on my door, you are not going to be going home."

"It was no longer fun for them, because I have a feeling that expressing fear or upset is the goal for most of them. Sure, this does not weed out someone who seriously wants to hurt me, but it puts the brakes on the sheer number of these things I get."

Confrontation

Most women are told not to respond to their harassers online. As Mikki suggested, trolls feed off our emotionality. Most safety experts say it is unwise to engage, that a harasser could escalate. Since social media platforms are built to amplify outrage, some online safety experts told me that continued engagement with a harasser could even signal the algorithm to elevate an abusive situation. Psychologists stress that shaming another individual is rarely effective at deterring them from bad behavior.

It seems reasonable and safe to suggest that women not try to confront, shame, humiliate, or intimidate their harassers online. Yet some women do, and many have told me they have experienced both practical and psychological benefits. Speaking directly back to a harasser can also qualify as a form of counterspeech, and as The Dangerous Speech Project's Cathy Buerger told me, not all counterspeech is civil. While speaking in an uncivil way is unlikely to change the original speaker's beliefs and runs the risk of normalizing uncivil speech online, she said it's also true that this kind of speech could be "a useful tactic to get something to go viral or to reach the larger reading audience."

Blanket suggestions to never respond to abuse may ignore women's unique vulnerabilities, their culturally specific ways of coping with antagonism and threat, their relationship to likability and respectability politics. It may also underestimate the feelings of empowerment associated with confrontative coping.

As Nina Jankowicz wrote in *How to Be a Woman Online:* "Women are expected to stoically endure astronomical levels of abuse to simply participate in conversations while navigating a set of social mores and boundaries that simply don't exist for men. When men encounter behavior they don't like online, they curse. . . . They willingly and openly dogpile and troll. And the world thinks them more manly for it." Women are expected to keep their mouths shut.

Robyn Mallett, a professor of psychology at Loyola University in Chicago, has found that women who ignore sexual harassment—which she defines as not responding—tend to become more tolerant of it. Mallett, who has also studied how personal goals influence people's responses to discrimination, has noted that when the goal to be respected outweighs the goal to be liked, women are more likely to confront harassment.

On my Twitter feed, I see many women addressing their abuse in ways that forgo likability. I watched as civil rights attorney and transgender activist Alejandra Caraballo told right-wing commentator Tim Pool to "eat shit" after he responded to her tweet on anti-LGBTQ extremists attacking parents outside a California school meeting during Pride. When a blue-verified user tweeted, "We will find you," she responded, "And do what exactly? I'm right here in Cambridge motherfucker."

Morgan Sung, who frequently uses her social media platforms to call attention to the suppression of marginalized voices, responds sardonically to sexist remarks. On TikTok, when someone left a comment on one of her videos that read, "Kinda . . . Basically . . . Sorta . . . That's what passes for modern journalism. A bunch of conjectures strung together to justify an opinion presented as facts," she responded, "Oh my god, you are so boring and this is the most boomer take. If you hate the sound of women talking that much then just read the article." In 2021 she responded to Dinesh D'Souza's tweet promoting a segment on Fox News, calling him a "snowflake ass."

Reproductive justice advocate Dr. Leah Torres snapped at the user who asked her if she heard the babies scream in her nightmares.

Women who fight back online are signaling a crucial refusal.

"I know why that advice is there not to engage," Mikki said. "But I really believe after all of these years of being online, that the people who are the most dedicated to being awful on the Internet are getting off on the fear and the silence."

* * *

In the absence of data, it is impossible to draw conclusions about which women are more likely to fight and specifically to confront, and which are more likely to ignore. Mikki told me she suspects race is a factor. She has observed differences in the way Black women cope with abuse online, and the way white women do. "Watching the advice white women give each other about online harassment, I'm very confused because it's very 'ignore the trolls,'" she said. "That doesn't work."

Several Black women and people of color told me they confront their harassers online because if they don't try to stop the abuse themselves, they are convinced no one else will. KáLyn Coghill, digital director of me too. International and an adjunct professor at Virginia Commonwealth University who studies how digital misogynoir impacts Black women and nonbinary people, is a frequent user of TikTok, Twitter, and Instagram. They told me their experiences with reporting have been outrageously unproductive.

"I have reported so many pages that are blatantly racist, blatantly misogynistic, blatantly sexist, homophobic, transphobic, and very violent towards Black women. And I've gotten responses from Twitter and Instagram and even TikTok saying things like, 'Oh, we didn't find that they were violating any of our guidelines.' "

Black women have rarely been able to depend on people outside their communities or in positions of authority to protect them. Black women get the benefit of the doubt less, are treated as victims less, get less support, and thus learn to expect less support. As scholar Christina Sharpe wrote in *Ordinary Notes,* "There is no set of years in which to be born Black and woman would not be met with violence."

Sydette Harry told me that for Black women, survival has always meant a degree of self-reliance. "I always say the cavalry isn't coming, and even if they came, I'm more likely to be their target than the person they're helping."

Confrontation is not innate to Black women, as many Black women engage in the taxing labor of suppressing their emotions daily. But fighting back and speaking up are sometimes required for Black women to be permitted a voice in public discourse. In her book *Check It While I Wreck It,* Syracuse University professor Gwendolyn Pough writes that the concept of "wreck," which has its roots in hip-hop, involves "the things Blacks have had to do in order to obtain and maintain a presence in the larger public sphere, namely, fight hard and bring attention to their skill and right" to be there.

This insistence can sometimes mean a renunciation of likability, though some Black women, like Mikki, do not believe they ever had it in the first place. Mikki told me she is unconcerned with being widely liked. Black and white women both navigate the likability trap, deal with different socializations around it, and weigh different factors in deciding when to speak, as well as what words to use. Being disliked can come with a cost for any woman, but white women may count on likability for safety in ways Black women cannot.

Philosopher Shiloh Whitney told me that as a white woman she often feels an "incredible resistance" to what she calls "breaking the social fabric." She finds it hard to even picture herself behaving confrontationally.

To illustrate her point, Whitney told me a story. She was on a New York City subway platform when a Black woman started yelling at a white man who was harassing her. As Whitney watched, people moved away, trying to separate themselves from the ruckus. She sensed that people on the platform felt the woman yelling was the problem, that her confrontation was more worthy of criticism than the man's harassing behavior.

But Whitney also noticed something else. "It achieved exactly what she needed it to, which is that the guy didn't want the kind of heat that she was bringing to the situation by screaming about this. . . . And so she got rid of him very effectively."

Despite the strategy's effectiveness, Whitney said she could still feel herself struggling to imagine behaving that way. She thought, as a white woman, she might not even have to, as it's possible that folks would have seen her being harassed on the platform and thought, "Oh, look at this nice white lady who's in trouble and needs help."

After my conversation with Whitney, I thought about my own relationship to confrontation. I have never been in a physical fight. I have never responded directly to a harasser online. I did not confront the boss who sexually harassed me at work. He owned the company, and I didn't even know where to file a report, so I quit instead. I have been stunned into silence by racist and sexist remarks. The only time I fought back was when a man reached under my skirt in a bar. I wonder how many worthwhile battles I have not fought, because I was preoccupied with keeping things calm and making things nice. I wonder how many times I have sacrificed another woman's safety, because I was invested in maintaining my own.

Community

The strength of a woman's community can be a decisive factor in whether she responds to violence.

In 2009 Stacey Ferguson founded Blogalicious, an annual conference that for nine years brought together multicultural women influencers. Even before social media, she told me, women of color bloggers would fight back in their comment sections. The blogger was the moderator, and she could decide which comments remained posted. Ferguson said rather than moderate out a hateful comment, many women would let the comment through and either reply directly or let their community do it for them.

Mikki told me when she responds to a harasser, other members of #BlackTwitter often join in, too.

A quarter of Black adults use Twitter, and as scholar André Brock wrote in a column for MSNBC, many of them participate in #BlackTwitter, a vibrant counterpublic where users engage in everything from political discussion to the cultivation of Black joy. Black Twitter is also a site for tech critique, as many Black digital feminists use the space to document and expose the harms of digital technology. KáLyn Coghill told me #BlackTwitter is a place where "you can show up as your full Black self." It has been an especially important space for Black women, whose stories so frequently go untold or become subsumed into Black men's histories. On #BlackTwitter, women share more complete narratives about who they are.

"We have been depicted on syrup jars as a mammy, and we're the welfare queen, we're the Jezebel, we're angry. We're all these things from the beginning of media representation all the way through now. And that's really not our story," said Brooklyne Gipson, an assistant professor of journalism and media studies at Rutgers School of Communication and Information who has studied Black feminism online. For a long time, she told me, "there was no popular place to have this conversation."

#BlackTwitter has enabled vital connections and the formation of deep social bonds. Black women in these spaces are not only speaking about the violence they face, they are creating and maintaining a culture of support and care.

When former *Ebony* editor Jamilah Lemieux was targeted by conservative trolls, it was other Black women online who had her back. The same for historian Anthea Butler and sociologist Eve L. Ewing.

But I also think back to Alexandria Onuoha, a young Black woman at a liberal arts college who hid her anger and suppressed her emotions so that she could continue to navigate a predominantly white space. Fighting back is different when you're a Black woman fighting with half a million followers, and when you're a student and the people you're "fighting" are your white professors.

Predictably, there is also a trap for women who fight back. One of the more consistent and robust findings from research on confrontation shows that bystanders tend to evaluate confronters negatively, especially when they use harsh or aggressive tactics. Women are expected to be perfect victims. Women must always be virtuous and nonretaliatory.

Emma Katz, an expert on domestic violence, wrote in a 2022 blog post that fighting back is not an "issue of morality. It doesn't say anything about the survivor's worthiness or their general character. . . . It is vital to remember that the survivor would have nothing to fight back against if the abuser stopped abusing them."

The refusal to meaningfully address women's online abuse means we all simmer in this toxic soup. I have often wondered how this may be changing our communication patterns. Is it easy to flip from humiliating a harasser to denouncing an intra-community violation? Sometimes women who fight back online are accused of hypocrisy, of engaging in the very tactics they critique. Where is the line between defense and retaliation? Between exposing harm and perpetrating it? Is it always clear who is punching up and who is punching down?

These are difficult distinctions. After comedian Maria DeCotis posted magakyle's dick pics and DMs on Twitter, her followers went after him. During his own onslaught, he asked her to take the pictures down. He messaged her, "Let's just forget this happened."

I asked her what she thought when she read that.

"I was like, 'Good. I'm glad my followers came after you.' I didn't ask them to do anything. I would never ask them to put any of their emotional energy toward someone who doesn't deserve it. But I posted it, and they went after him, and I was like 'Good. I hope that they say more mean shit to you, and I hope that you have to feel how I'm feeling right now, even though you never will.'"

In her book *This Is Why We Can't Have Nice Things,* scholar Whitney Phillips asks: "If the goal is to dismantle patriarchal structures, and if feminist trolling helps accomplish those ends, then are the means, however problematic, retroactively justified?"

She leaves it for the reader to contemplate.

For women who have been ignored and bullied and harassed for so long, it should come as no surprise that some may cheer or even participate when they see a reversal, when what happens to them all the time happens to someone whose own speech contributes to their oppression. When belonging and identity and personhood are questioned, challenged, attacked, it makes people all the things you would imagine: anxious, scared, angry, contemptuous.

It is important to acknowledge that there are differences between Black women networked online trying to shut down a white comedian perpetrating misogynoir, and a white man sending his followers to pile on a young female reporter. I also think we cannot be surprised when our values do not always precisely line up with our own behaviors, especially when we are mired in the mess of trying to survive these abusive spaces.

While I was working on this chapter, a friend sent me an essay from queer poet Andrea Gibson on how some activists are muting their own voices because of nonstop criticism online. I am sharing an excerpt here because I believe online abuse has coarsened all discourse, and as Alice Marwick's research on morally motivated harassment has shown, some of us are being harassed by people from our own social and political groups. Gibson said we cannot "afford to lose more voices in the fight against fascism." They encourage us to ask:

"Is your desire to publicly point out another's poor choice fueled in any way by an unwillingness to address and account for your own behaviors? Do you feel pleasure or excitement when your community piles on to further shame the person? Are you in any way overstating harm? Is toxic individualism impacting your capacity to root for and believe in the growth of others? Do you have a keen eye on the fact that particular brands of public callouts for even small judgment errors in today's world can mean the end of a person's employment, which commonly means the end of their health benefits and housing? Such questions are important because I'm not writing this essay about the alt-right. I'm speaking about community members who are committed to bettering the world, and are, like everyone, making mistakes on that journey."

Empowerment

Some research has found that women who confront sexism experience increased feelings of empowerment, competence, and self-esteem. Confrontation can have a positive impact on mental health, and it can prove the more beneficial strategy over time.

Mindi Foster, an associate professor of psychology at Wilfrid Laurier University who studies how women cope with gender discrimination, conducted a study where for twenty-eight days women kept track of daily experiences of discrimination in an online diary. They described how they responded to the experiences, and they completed measures of well-being. Results showed that at the beginning of the study, the women who used indirect confrontation—signaling that something was not okay (storming out, rolling their eyes, using sarcastic humor)—had greater well-being than those who used angered confrontation, expressing their anger and beliefs about the unfairness of the incident (getting into an argument, telling someone they had no right to say that). But as the month wore on, the women who continued to use indirect confrontation saw their well-being decrease, and the women who continued to use angered confrontation saw their well-being increase.

PEN America doesn't encourage counterattacking harassers with violent language, and it underscores that employers may have strict social media policies around engagement. But its experts recognize the importance of strategies for managing online abuse that facilitate women's empowerment. In an online abuse defense training session, PEN America experts said that there can be value in combating lies about yourself and fact-checking false claims about your work, though from the standpoint of best practices and guidance, they say it is rarely productive to speak directly to a harasser.

"Every now and then, there are instances where the harasser and the person being harassed are able to have a constructive and empathetic conversation," PEN America's Viktorya Vilk told me, "but often it leads to escalation and further abuse, to wasted time and energy."

When a woman is considering whether and how to fight back, PEN America suggests that she ask herself if she is emotionally or psychologically ready for a potential confrontation or escalation of abuse, which may happen. She might ask herself, "Would saying something feel empowering? Do I feel it's worth it to speak directly to the harasser, given the possibility of confrontation or escalation, or can I speak out more broadly about my experiences or my work? How can I use my voice in a way that feels safe to me?"

Mexico-based psychoanalyst Ana Zellhuber, who collaborated with the International Women's Media Foundation on a mental health guide for women journalists facing violence online, discourages women from responding to harassment. But if a woman feels a retort is necessary, she said, she should first assess the risk. Women have different risk profiles depending on where in the world they live, who is targeting them, and what else in their life makes them vulnerable.

There's a difference, Zellhuber said, between a troll attacking you, a man you know attacking you, a government attacking you, and organized crime attacking you. For some women, responding to an online abuser could put them at more risk for offline violence. Some harassers may back down when antagonized, while others are eager for attention and grow energized through engagement.

These calculations are exhausting. In her article on "feminist digilantism," UNSW Sydney associate professor Emma A. Jane wrote that "do-it-yourself strategies" have some benefits, but alone are not "an adequate solution to the broader problem of gendered vitriol online," because they shift "the burden of responsibility for the problem of gendered cyber-hate from perpetrators to targets, and from the public to the private sphere."

Journalist Morgan Sung told me she finds herself internalizing this tension: "I do respond to trolls and I do fight back, because in the moment I'm like 'I'm mad and I'm going to win this argument. And I'm going to give myself that little win.' But then I think there's also that voice in the back of my mind being like 'Is this helping? Is this going to stop the harassment? Is it going to make any of this better?'"

Many of the women I interviewed for this book did not respond directly to their harassers but fought back in other ways. They were researchers who studied women's responses to harassment, activists working with platforms to change harmful products and policies, and legal experts using litigation to create accountability for digital harms.

I often thought about what I might face when these words reached an audience. While most of my interviews centered my subjects, occasionally my subjects would check in on me, and almost all who did were women of color. They said, It's a long fight, we need you. They asked, Are you okay? Are you resting, are you taking breaks? Are you still finding joy? It is unending work for these women to build supportive communities. I doubt I've done enough to deserve such an extension of care.

Halfway through reporting this book, one of my interviewees said something to me that I didn't know I needed to hear. She offered the protection and the prediction I needed, so I would not be stunned into silence.

"You are going to be attacked, and don't worry, because we're here for you," said AI technologist Mutale Nkonde. "They're going to attack you, and then we're going to bite back."

CHAPTER 13

A Humanity That Isn't Up for Debate

IN HER FIRST memory of Johnny Depp, Rebecca is five years old. She's watching a commercial for *Edward Scissorhands,* in which Depp plays an artificial man who falls in love with an ordinary girl. He is mostly a man, apart from his unfinished hands, which are fashioned of blades. The commercial features a scene where one of Edward's scissor hands punctures a waterbed. A small fountain springs up, which he clumsily plugs with a quilted plush dog. Rebecca's parents had a waterbed, and they were always worried that she would destroy it—that the wrong toy, the wrong writing utensil, would trigger a deluge. Young Rebecca had internalized the fear. The scene stuck.

The tagline on the 1990 movie poster for *Edward Scissorhands:* "an uncommonly gentle man."

Rebecca didn't pay much attention to Depp after that. She was a kindergartner, after all. But in high school, she finally watched the film in an English class: a study in aesthetics. She was captivated by the haunting atmosphere, the gothic style. She watched *Sleepy Hollow* and loved that, too. Then *Pirates of the Caribbean* arrived, and Captain Jack Sparrow's androgynous charisma tickled Rebecca's teenage heart, thickening an illusion of intimacy that solidified a parasocial relationship with the actor that Rebecca would not escape for years. Stores started selling "I Love Johnny Depp" necklaces. Of course, Rebecca bought one.

Rebecca began watching all of Depp's older films. In an act of devotion, the person she was dating in high school spent $300 on a copy of the *Pirates of the Caribbean* script, signed by Keira Knightley, Orlando Bloom, Geoffrey Rush, and Depp himself. Rebecca treasured it. She took the script with her when she left her parents' house and hung it up in her first apartment.

But in 2016 actress Amber Heard, who just fifteen months earlier had married Depp, said the actor had hurt her and she was leaving their marriage so he could not hurt her again. She filed for divorce and then a restraining order. There are pictures of her from the courthouse, a small bruise on her right cheek. She looks so tired.

Rebecca had exalted Depp, and given the human reluctance to relinquish our heroes, it would not have been surprising if she had dug in her heels, insisting on his innocence. But she did not. Rebecca had loved Johnny Depp, which meant she had read everything written about him. She read the reports that he assaulted a security guard in 1989 and that in 1994 he was responsible for thousands of dollars' worth of damages to a hotel room he was staying in with then girlfriend Kate Moss. She had read about his struggles with alcohol and drugs.

In Rebecca's mind, Heard's accusations tracked. She believed her, and she was willing to say it out loud. "I felt like he was a dick bag who lied to me, and I wasn't going to take it sitting down," she told me.

Rebecca said she started "yelling at people" on Facebook who defended Depp. She yelled at people on Facebook during the UK libel trial. Depp lost. A London judge found the words of a British newspaper calling Depp a "wife beater" to be "substantially true."

People said hateful things to Rebecca online. They said hateful things to anyone who tried to support Heard. Just as the U.S. trial began, Rebecca decided to start a Twitter account to document the harassment of Heard's supporters. She called her Twitter account @LeaveHeardAlone. She thought, whether Depp was defamed or not, the harassment people were experiencing for supporting Heard was unacceptable. A couple of days before she started tweeting, she received some Facebook messages telling her she should be gang-raped and cut up. She decided not to use her real name on social media anymore. In these pages, she will go only as Rebecca.

Rebecca watched and listened to the trial without commentary. She listened as Amber Heard testified to horrific details of abuse, listened to evidence that Depp sexually assaulted her with a bottle, beat her, berated her, controlled her, put cigarettes out on her. Cameras in the courtroom turned the trial into a spectacle. Heard was a young bisexual woman who worked consistently but hadn't become a household name. Johnny Depp was one of Hollywood's biggest stars. To experts in domestic violence, the power dynamic seemed clear, but the public didn't see, sometimes couldn't see, past Heard's human responses to violence. Heard's legal team was also denied the opportunity to present evidence that would have helped clarify what remained frustratingly muddled—evidence such as therapy notes that recounted violence during the relationship and text messages from a Depp employee that appeared to corroborate Heard's claim that he kicked her on a private plane.

The media did little to help, putting out stories with headlines like "mutual abuse," a term used during the testimony of the couple's former therapist, though experts in intimate partner violence say the term is not used in the field, because abuse is never mutual: there is always an instigator and someone who responds to the violence, there is always an imbalance of power. Many journalists did not or could not contextualize what was happening in the courtroom, because that would have required an understanding of the dynamics of intimate partner violence and a willingness to expose them.

On social media, Amber Heard became a punchline. People online treated her as if she couldn't feel pain. Because they didn't believe her, or because they didn't care, they decided she didn't matter.

She said she was physically, sexually, and emotionally abused. The Internet laughed, and it's possible the jurors, who were not sequestered, watched. Perhaps they laughed, too.

Rebecca was stunned. She wondered why others couldn't see what she saw. She told me she wondered if people were insane.

She remembered the O. J. Simpson trial over the murder of Nicole Brown Simpson. It was radical feminist Andrea Dworkin who said, "You won't ever know the worst that happened to Nicole Brown Simpson in her marriage, because she is dead and cannot tell you. And if she were alive, remember, you wouldn't believe her."

Rebecca remembered Monica Lewinsky. She still feels guilty for being an "asshole" about Monica Lewinsky. (She was thirteen when the Clinton-Lewinsky story broke.) "I'm sitting here going, 'Y'all, it's happening again.' . . . And I just decided that I was going to do something about it."

Rebecca was part of a small group of Twitter accounts publicly supporting Amber Heard. She was up against not only dangerous misperceptions about domestic abuse and zealous Depp fans but also venomous online harassment that was both organic and significantly coordinated. During the defamation trial, jurors heard one of Depp's lawyers say he communicated "episodically" with what he called "Internet journalists" and specifically admitted to communicating with an online creator who made anti-Heard content. Rebecca was up against a mass of influencers who realized they could monetize misogyny to grow their platforms and line their pockets. The truth was ancillary.

Comments on the live stream videos of the trial were so awful, Rebecca said, that someone suggested she start a Discord server—a private chat room run by a volunteer administrator who decides membership and sets rules—where people could watch the trial in a supportive environment. Shortly after starting the Discord, Rebecca's Twitter followers ballooned.

Rebecca suspects she filled a gap. The few other accounts supporting Amber Heard provided facts and context, but Rebecca was focused on something else.

"We need to make the community," she thought. "We need to make a collective."

Social Support

I think about where this book began: Gloria Steinem's apartment, in 2017. Since the 1960s, Steinem has considered the space not only her own home but the home of a movement. When I spoke with her last, over Zoom, I told her that her home was a site of inspiration for me, given our last conversation there.

She was visibly pleased. "That's why it's so important that we say what we're thinking," she told me. "We have no idea what single word may be useful."

The women I interviewed for this book used various strategies to cope with violence, but there was a single component of care almost every woman said was crucial to their survival: other people.

Coping may be an individual's capacity to soothe and regulate themselves, but abundant evidence shows that one of the most important aspects of self-care is developing a mutually supportive community.

The philosopher, the neuroscientist, the psychologist, the sociologist, the digital scholar all agreed that externalizing painful experiences of abuse with like-minded or compassionate others is among the most beneficial coping strategies.

As Dr. Thema Bryant writes in *Homecoming:* "You deserve a community where your humanity is not up for debate."

Study after study shows that social support, the feeling that we are loved, cared for, and valued by other people, is a key component of psychological and physical health. Social support acts as a buffer against bad outcomes. It makes us feel validated and less alone. Social support can make us more resilient and less likely to develop PTSD after a traumatic event. There is a reason why peer support—an evidence-based practice used in veterans' groups, in suicide prevention, in substance abuse prevention—is being used as a response to online violence.

As psychologist Emily Sachs told me: "Social support . . . can counter the stressor by reminding ourselves of who we are, that we do belong in community, that we have value, that we're not damaged goods, that the world has good and justice and love in it as well as toxic elements, that we may be safer than our bodies feel, that people will show up."

Rebecca said she credits the success of her account, which has grown to over thirteen thousand followers on Twitter, to her focus on forming a community of people who could support one another through the trial, as well as work together to counter misinformation about the case and survivorship broadly. She said she came to need the community as much as it needed her.

Rebecca was doxed on a pro–Johnny Depp forum on Reddit. She was abused directly. People still screenshot her tweets to disparage her. Someone made a "LeaveTurdAlone" parody account. When Rebecca raised money for domestic violence organizations in Heard's honor, people online said she was stealing the money for herself, grifting, lying. She watched as they made up their own stories, said whatever they wanted, no matter what was true. She watched as people who supported Heard grew increasingly afraid. There was an important pro-Heard Twitter account run by a woman who went by "Kamilla" that Depp fans managed to bully off the platform completely. "I have to wake up in the morning and steel myself to whatever's going to show up in my inbox," Rebecca told me.

During the worst of it, friends would check in each morning, ready to report credible threats, encourage her, ask her what she needed, ask how they could help.

When I asked Rebecca what kinds of supportive messages she

received from her core group, she struggled to identify them. It wasn't necessarily anyone saying "keep going" or "you're doing great," she told me. Often it was just the existence of a space where she and like-minded others could say privately what they would never say publicly.

"Sometimes the core group was making fun of Deppie bullshit, or it was venting about some of the worst and most offensive Depp fans," she said.

It was the energy of the space—the Discord server, in particular—where she could commiserate about the trial or about life, where she could "be a little more human and messy and not so anonymous," she said.

Deleting hate, muting it, blocking it, dismissing it, reporting it, tucking it away, in a file, in your body, in your mind, will only ever get you so far. The goal of online abuse is to isolate us. To fight these tactics, we need one another. As Dr. Judith Herman writes in *Truth and Repair,* "People cannot feel safe alone, and they cannot mourn and make meaning alone."

Almost every woman I interviewed for this book valued a support system of other people they could talk to, who listened, who refused to look away: people who intimately understood what they were experiencing or at the very least tried. Sportswriter Nancy Armour called it her "girl gang." Journalist Taylor Lorenz has a group text with other women on her beat. Representative Leigh Finke said Representative Ilhan Omar and Lieutenant Governor Peggy Flanagan helped make her first year in the Minnesota legislature survivable. When Alexandria Onuoha was struggling in the aftermath of her op-ed, one of her professors helped her through her final year of school. Plumber Brooke Nichols has the other "baddies" in the trades. Cultural critic Mikki Kendall has #BlackTwitter.

Neuropsychologist Negar Fani said sharing experiences of social pain with an empathetic other can actually reduce the pain itself. Social support can help alleviate distress, clarify our experiences, and aid us in the process of reporting, whether to social media companies or to law enforcement.

Friends, especially, can offer deep and sustaining forms of empathy. UCLA social psychologist Naomi Eisenberger found that when a person sees a stranger experience social rejection, they don't, on average, show pain-related activation in their brain. They show activation in regions involved in mentalizing—thinking about the minds of others, thinking about what they must be thinking or feeling—but not pain-related activation. But when a person watches a friend experience social rejection, then they too feel pain.

We need people, especially those who love us. We need people who pick up the phone, who send the text, who make the joke no one else could. We need people who will listen when we ask, "I'm not the crazy one, right?" And tell us, "Right, you're not the crazy one." Sometimes we need advice and practical support, and sometimes we need people who can simply remind us of what we are doing and who we are.

When I was reeling from Trump Jr.'s mob, my editor reminded me: "I'm proud of how you tackle really hard things and try to make the lives of victims better."

I read this in my kitchen the second night of the onslaught. I sat on a stool while my family danced. The toddler naked, her belly a river of melting pop. Bumping butts and grunts and giggles. My oldest played a song on her keyboard. For the first time in forty-eight hours, I could hear more than fists on the door. I could hear a melody.

Solidarity

During my interviews, it was clear that social support connected to another crucial concept: solidarity. Solidarity refers to an individual's commitment to working with others in a larger transformative struggle. Critical justice scholar Loretta Pyles said social support is individualistic, it is about what an individual needs to cope with stress or a crisis. Solidarity is a different paradigm that values the collective, though there are important relationships between the two. Psychologist Deborah Plummer told me that "social support is the bedrock for how change happens in our communities. . . . If my core isn't speaking to your core as a human," then we cannot come together in solidarity.

Solidarity, support, and community are reasons people flocked to the Internet in the first place. It is why many women remain. As Bridget Todd said on an episode of her podcast *There Are No Girls on the Internet,* "One thing I love about being somebody who is a woman online is people will ride for you, people will fight for you, they'll speak up for you, they'll shake tables for you, they'll make sure your voice is included and amplified."

Women can show one another they are worth fighting for. And men can show us that they, too, are interested in ushering in a different kind of world, one that does not so easily occlude women's voices, that invites debate and disagreement but refuses to accept abuse and derogation as the price of participation in public and civic life. The current cost is one many of us cannot pay, so many of us will simply not speak.

When comedian Kelsey Caine was destroyed online over her caricature of Louis C.K., male comedian Adam Conover retweeted her parody and followed her on Twitter. Sportswriter Nancy Armour said that during some of the worst of her harassment, it was a male editor who had her back.

As I pushed for this book to exist, it was a male colleague who counseled me through the process, offering constant encouragement and practical advice. When I was trying to sell this book, when I faced rejection, when I cried at my desk, in the bathroom, on the hot driveway, it was my husband who told me: "You can do this. You will."

Standing in solidarity is not always easy. As Sara Ahmed writes in *Living a Feminist Life:* "When we have to fight for an existence, it can shape our encounters with each other. There is no doubt we can experience each other as sharp and brittle. We come up against each other." Solidarity requires humility, introspection, a willingness to see when we might be the problem.

While writing this book, I was struck by women's eagerness to engage with me, without trust necessarily established, knowing there could be a cost. Women made themselves available on deadline, while writing dissertations, on vacations, on weekends, juggling work, juggling children.

When I noticed holes in my reporting on Marissa Indoe's story, I reached out to Taima Moeke-Pickering, a professor in the School of Indigenous Relations at Laurentian University, and she got back to me almost immediately, making time to talk over Zoom. When I thanked her for agreeing to speak over the weekend on such short notice, she responded matter-of-factly:

"I will always make time for a sister in the fight."

"A Global Gaslighting Event"

The defamation trial between Johnny Depp and Amber Heard took on a cultural significance that most people were unprepared for. It showed the dangers that women continue to face in a patriarchal culture, and it exposed the power that unchecked white supremacy and misogyny online have to shape offline outcomes.

The trial showed how deeply we misunderstand violence, especially the dynamics of intimate partner abuse. The talking points of men's rights activists went mainstream, claiming that men are under attack by lying women, that false rape accusations are everywhere. We saw disinformation spread for influence and profit and become misinformation that women survivors of domestic violence used to protect a man they admired, to discredit a woman who they believed could not be a victim because she did not look and act as many of them did.

While Gamergate showed how organized, wide-scale harassment campaigns could cause harm, Alejandra Caraballo, an attorney at Harvard's Cyberlaw Clinic, said platforms did not consider the financial incentives that could go along with it: from creators monetizing videos mocking and discrediting Amber Heard to those selling "Justice for Johnny" T-shirts, we witnessed "the professionalization of the harassment campaign," she said.

These campaigns can be hugely profitable. The analytics service Bot Sentinel, which investigated Heard's harassment and called it "one of the worst cases of cyberbullying and cyberstalking by a group of Twitter accounts," has also tracked the targeted harassment of Harry and Meghan, the Duke and Duchess of Sussex, showing how "single-purpose hate accounts" have "turned targeted harassment and coordinated hate campaigns into a lucrative hate-for-profit enterprise." Bot Sentinel estimated that three anti–Harry and Meghan YouTube channels raked in hundreds of thousands of combined earnings.

YouTubers who would see a couple of thousand views a day on their channel suddenly started seeing a million views in that same time period after switching to covering Heard. NBC reporter Kat Tenbarge wrote about a fifteen-year-old boy who had been getting a few hundred views for his YouTube videos about the game Elden Ring. When he decided to make a video about the Depp-Heard trial, within a week his new content had more than 10 million views.

These pivots contributed to what Tenbarge called "a global gaslighting event."

People who asked neutral questions about Amber Heard in Johnny Depp subreddits were doxed. People who tried to express support for the actress on social media were threatened. Journalists like myself who tried to write measured content were subject to an avalanche of hate.

Women are often told their abuse is the cost of free speech. But here was a woman who tried to speak. What happened? Depp sued

Heard for $50 million, claiming she defamed him in a 2018 *Washington Post* op-ed where she referred to herself as "a public figure representing domestic abuse." Depp won his defamation suit, and the case resulted in Heard's "global humiliation," which is exactly what Depp had promised would happen in a 2016 text after she secured her restraining order.

"I am harassed, humiliated, threatened, every single day," Heard said on the stand. "People want to kill me, and they tell me so every day. People want to put my baby in the microwave, and they tell me that."

Many women wanted to stand with Heard but could not. Many of the women who wanted to stand with her were survivors of violence themselves, and they were exhausted and afraid. The harassment did precisely what it was designed to do.

It *silenced.*

Tenbarge was one of the rare mainstream journalists who tweeted her observations and analysis about the case in ways that corrected misperceptions about abuse and challenged the omnipresent narrative that Depp was innocent. When she decided to tweet about the trial, she was prepared for backlash.

She told herself: "I'm going to say the truth about this. I'm going to say what I observed about this, and I know it's going to end badly for me, and I'm going to do it anyway. . . . I myself have never been in an abusive intimate partner relationship, and I've never been sexually assaulted or raped. . . . I was like, I'm going to carry this burden of the hate and harassment that I'm going to encounter on behalf of these women who can't."

Rebecca was also willing to carry the burden, but it was her own history of sexual violence that animated her. She has been sexually assaulted multiple times. She knows what it feels like not to be believed.

When women experience violence online, they want to be believed. When they say they felt terror, when they say they were hurt, when they say it destroyed their life, when they say they can no longer do the work they want to, that they can no longer express what they need to, they want other people to trust these disclosures. They want people to respond, and not just with a privately supportive text or email. They want allies in demanding a transformation.

"The first thing that everybody says when they talk to somebody who has gone through this is 'I'm so sorry,'" disinformation analyst Sara Aniano told me. "Of course, that's what you say, right? But what I want to do is shake them and be like 'No, don't be sorry. Be very angry and do something.'"

The "something" could be a public show of solidarity. If that's too risky, it could be epistemological—reading literature and research on the issue. It could be using your purchasing power to help sustain a marginalized creator online.

As Dr. Judith Herman writes in *Trauma and Recovery:* "The victim . . . asks the bystander to share the burden of pain. The victim demands action."

The Mob

I've saved this story for the end because I didn't want to tell it. It lives in the parts of me that are most distorted, the parts that are never really sure. In this story, words matter, they hurt, they corrupt a body. In this story, misinformation spreads, outrage rules, pain is exploited. I suffered most when I felt disempowered and alone.

To understand what happened, I'll start with Peter.

Peter was seventy-five years old when he emailed me in the summer of 2021 and disclosed that he had been sexually abused at a Vermont summer camp when he was a boy. I wanted to explore his half century of suffering. For six months I conducted interviews, pored over fragile camp records, located other survivors, spoke with a descendant of the camp's founder, and reached the alleged perpetrator. I wrote Peter's story.

I paired the seven-thousand-word narrative with a short companion piece on the science of pedophilia, in the hopes that it might help readers better understand the perpetration that had contributed to Peter's suffering. Peter often ruminated about his perpetrator aloud. "Why did he do this?" he would ask. He wondered how to reconcile his abuser's harm with his humanity. I knew Peter wasn't the only survivor with these questions. The story sought to explain pedophilia and to debunk misperceptions that experts say contribute to child abuse.

I had one day of gratitude and relief: the day Peter's story published. He was proud, and I received dozens of emails from men who had also been abused when they were children, telling me they saw themselves in the story.

But the next day my piece on pedophilia, which had been behind a paywall, was shared in a Twitter thread on one of *USA Today*'s accounts. Immediately afterward it had hundreds of outraged retweets slamming *USA Today* for trying to "normalize" pedophilia. A Twitter account associated with QAnon directed people to attack me. The Southern Poverty Law Center describes QAnon as "a sprawling spiderweb of right-wing internet conspiracy theories with antisemitic and anti-LGBTQ elements that falsely claim the world is run by a secret cabal of pedophiles." Former president Donald Trump has indicated he believes them.

Editors said to protect me, they deleted the thread and posted a tweet stating that the original thread lacked context. They also removed the story from behind its paywall, hoping people would read the piece in full rather than form opinions based on a headline alone.

They did not say, "We stand by our reporter."

Shortly after the thread was taken down, Donald Trump, Jr., tweeted a screenshot of it, lambasted *USA Today* for trying to "UNDERSTAND" pedophiles, and accused *USA Today* of trying to hide its missteps by deleting the tweets. Then he pinned his tweet to the top of his Twitter page, where it remained for several days. News stories assailing me appeared on Fox News, Breitbart, and the *Daily Mail*.

USA Today supported me practically, and many editors and colleagues offered emotional support. An editor on the social media team took the keys to my accounts and worked with the highest levels of Twitter to keep me safe. I received many texts and phone calls checking in to make sure I was okay.

I did not, however, have agency around the decisions made in response to the harassment, including removing the original Twitter thread and changing the headline on the story, and I did not have public institutional support, the action I repeatedly asked for.

During the reporting of this book, I read a study on online harassment of women scholars titled "I Get By with a Little Help from My Friends." Researchers approached their interviews using an ecological model, which considers multiple influences that shape human behavior, and found that during and after an experience of online abuse, women depend on three levels of support: personal and social, which includes encouragement from friends and family; organizational, technological, and sectoral, which includes support from an employer, a university, law enforcement, and the tech platform itself; and cultural and social attitudes, which involve the way we think and talk about gendered harassment or the online/offline divide.

Shandell Houlden, co-author of the study and a postdoctoral researcher at Royal Roads University in British Columbia, said while "in neoliberal individualized cultures the responsibility for problems lands on the individual," an ecological model of understanding helps clarify that victims need support from all levels; it encourages institutions to understand how social media operates and how it's being manipulated to cause individual and societal harm.

When I was mobbed online, I felt some layers of support and the limits of others. I believe my institution missed an opportunity to censure my harassers, many of whom are intent on discrediting mainstream journalism and shutting down women journalists, and it failed to push back against the dangerous Internet subcultures involved in the campaign that exploit people's moral outrage over the sexual abuse of young children. Decisions were made in response to reactions from conspiracy theorists, members of the far right, and the son of a president who sanctioned and benefited from white supremacy.

In its *Online Harassment Field Manual,* PEN America recommends that employers consider issuing a statement of support when an employee is being attacked for their work, if the targeted employee would find that empowering, as "the power dynamics between a lone target and an abusive (often coordinated) mob are extraordinarily uneven" and a public statement lets "staff know you have their backs by taking a stand against hate and harassment online."

It would be arrogant to suggest there were no legitimate critiques to be made of the story. Reporters are obligated to listen to readers, to take seriously the responsibility that comes with the power of a byline. But I can't remember anything in my inbox resembling critique. If there was thoughtful, concerned analysis online, I couldn't find it among the litany of threats against myself and my children. It was also evident many people were not reading my words but riding the outrage.

When Trump Jr. pilloried *USA Today* for daring to "'UNDERSTAND' PEDOPHILESS!!!" I said to anyone who would listen: What is wrong with trying to better understand violence?

Editors told me not to respond publicly. The story remains on the site without a correction.

I tried to convince myself that my newsroom did the best they could. I tried to tell myself a story I could live with, because I wanted to stay in my job. A year and a half later I called a former colleague to refresh my memory on the harassment. I sat in a Trader Joe's parking lot and began asking questions, but I didn't initially let her answer, because once I started, I couldn't stop, and I heard her somewhere in the middle of my rant gently encourage, "Let it out."

I asked: Why didn't anyone publicly defend me?

She said she didn't know. I called someone she had reported to, a more senior editor, who also listened to me for nearly an hour, as I implored her for an answer. When I asked her during the attack why no one publicly defended me, she told me it was to avoid provoking my antagonists. When I asked her this time, she said, "We should have."

• • •

Rebecca still has her autographed *Pirates of the Caribbean* script. It's in a box now, since she moved from her apartment into a home with her wife, but she kept it up on her wall even after Amber Heard came forward. She didn't want to erase her mistake. The script reminds her that in letting a hero die, a new kind of consciousness could be born.

Rebecca continues to structure her life around her harassment. It's so much work to stay psychologically and physically safe. She never talks about her platform with people she doesn't trust. When she is with family or friends and they ask her, What are you up to?, she tells the lie she uses to remain safe: Nothing.

Rebecca would like to disclose her experience if for no other reason than to boost her résumé. Her account has thousands of followers, and she's raised thousands of dollars for domestic violence organizations. Her research during her master's program was on activism frameworks, and she's considered using what she's learned about online movements to pursue a PhD. She can see herself doing public relations for a domestic violence organization.

"I'm kind of hoping that eventually I can be like, 'Hey, it's me.'"

When another big account supporting Amber Heard went offline, Rebecca's followers worried that she would, too. People frantically messaged her: *You're not going to do that, are you? You're not going to leave? You're not going anywhere? You're not going to stop?*

She told them: *No, I'm not. I'm not. I can't.*

And she knows the source of that strength.

"If I didn't have this group of people who were helping me and who I could talk to, I don't think I could still do this. I don't think I could still do it if it was just me."

Conclusion

I WONDER WHAT IT might take to rebuild the Internet with different values. What would it feel like if our digital spaces were sites of mutual responsibility, scenes of gratitude and reciprocity, rather than ones of dominance and competition? What if we traded self-promotion for productive humility? What is possible if we imagine that our most important value is not what we are allowed to say to one another but what we are obligated to do for one another? What would it look like to engage on social platforms whose goals were not production or profit but sustainability and healing?

There is nothing about our online lives that is inevitable. The Internet could have been many things. It still can be. It's an incredible tool, a testament to human innovation. It has helped raise consciousness, build movements, and save lives. It has also created vast amounts of individual, social, and political harm.

"The Internet is an artifact of human creation," Columbia University research scholar Susan McGregor told me. "It can be different. It is not that way because it must be. It's just that way because that's how humans built it. . . . We don't need to worry about what's possible. We need to worry about what we want. And we'll find a way to get to it."

According to the Pew Research Center, in 2017 Republicans and Democrats held similar views about the balance between free speech and online safety. But when Pew conducted its survey in 2020, partisan differences had grown fivefold. According to Pew, 60 percent of Democrats said it is more important for people to feel welcome and safe online than for people to be able to speak freely, while about 45 percent of Republicans said the same.

Civil rights attorney Alejandra Caraballo told me that a shared goal should be to maximize free speech, but not at the cost of individual dignity. People can criticize their governments, but they shouldn't be allowed to harass a person based on their shared identity or membership in a protected class.

Women and other marginalized people understand the value of free speech. We want to speak. We want to tell the truth about our lives. As philosophers María C. Lugones and Elizabeth V. Spelman have argued, "We can't separate lives from the accounts given of them; the articulation of our experience is part of our experience." We need to give our own accounts because "it matters to us what is said about us, who says it, and to whom it is said: having the opportunity to talk about one's life, to give an account of it, to interpret it, is integral to leading that life rather than being led through it."

• • •

There are paths to making a safer and more welcoming Internet.

PEN America's Viktorya Vilk said we need more training at every level on online abuse, self-defense, and institutional responses. Journalists specifically must be trained in how to avoid becoming purveyors of weaponized disinformation. PEN America is also pushing for better reporting policies, which it says are possible "with time, resources, and will."

Researcher Caroline Sinders said platforms need to grow their trust-and-safety teams, involve content moderators more meaningfully in those teams, quadruple those workforces, and ensure they are localized to the regions they're covering.

Tech investor Ellen Pao wrote in her book *Reset* that the entire tech system has "exclusion built into its design." Venture capital firms and technology companies must make it a priority to hire more women and people of color who can help develop ways to mitigate existing harms, as well as spot new ones before products are rolled out. Platforms must work to ensure their products are looked at and touched by a whole range of people. Pao, who was briefly CEO of Reddit before being harassed out of the job, made it a priority to reduce harm on the site, to remove revenge porn, to shut down some of the worst subforums. For her efforts, she was attacked by trolls. Eventually, she resigned. Researchers found that the changes she made led to an 80 percent drop in hate speech.

There is a complex policy debate around Section 230, a law that shields technology companies from liability for content produced on their platforms. Some legal experts say it is time to start thinking of platforms as publishers, denying them the immunity they have long enjoyed under Section 230. Treated as publishers, platforms can enact norms around acceptable forms of communication. They could be held liable for not enforcing them.

"Things do suck," lawyer Carrie Goldberg told me, "and that's really because these products are made to be abused, and the platforms were created without any consideration for abuse, but just as with any other product, we as human beings and consumers can require that there be certain minimum standards."

Others are wary of sweeping changes to Section 230, arguing it is a critical law for protecting the speech of ordinary people. Digital rights nonprofit Fight for the Future says that while Big Tech does abuse its power, removing or even reforming Section 230 could worsen censorship, which it says would disproportionately impact marginalized communities and silence social and racial justice movements. To hold Big Tech accountable for harms created by its products, it urges legislative action on privacy and civil rights as well as better enforcement of antitrust law.

Information scientist Jessica Vitak told me platforms must hold people who engage in problematic behaviors accountable, and not just with a slap on the wrist. For that to happen, Vitak said, there need to be policy changes, both within the platforms themselves and through federal legislation, that provide clear guidelines on consequences for engaging in harmful behavior online.

It behooves us to look more deeply at the efficacy and limits of counterspeech. As Bianca Cepollaro and co-authors argue in their paper on the practice, we do not yet have enough evidence to suggest that counterspeech is more effective at reducing speech harms than restrictions, and while we may not need to have a univocal answer on whether counterspeaking or restrictions are more effective—Cepollaro told me that each may prove the right solution in different situations—we need to investigate the assets and drawbacks of both tools in context. If we decide to rely on counterspeech to moderate the effects of harmful speech online, we need to understand how to use it in ways that are effective and reliable.

Journalist Koa Beck encourages users to remember they have power to pressure platforms to improve. "These are not town squares, these are not public spaces. . . . And I think somewhere in the narrative of social media and its power, that somehow got lost," she told me. "I look at social media as the next terrain of union activism, people walking off the platform, challenging rules. . . . It absolutely can be done."

Many scholars say it is essential to fund more research on the impacts of computer-mediated communication.

In previous eras of communication, we were able to observe over generations the effects of new technologies on adults and children, to determine how they aligned with our values and to put in place necessary safeguards. But the Internet has with stunning speed and force created a communication evolution, and we do not have nearly enough chronological data on how computer-mediated communication is changing our linguistic practices, how it is influencing our behavior, how it is affecting our brains.

"It has taken humans well over two hundred thousand years to evolve into our current form and only two hundred years for computers to leap from the analytical engine to artificial intelligence," scholar Claire Hardaker told me. The swiftness of transformations like these poses risks, and it is necessary to understand how new and often ubiquitous technologies are changing the texture, functioning, and experience of human life.

Online safety advocates Carol Scott and Melissa Eggleston told me that creating more trauma-informed technology can help us move in a more ethical direction. When we don't experience as much harm in the form of online abuse and exclusion, when we aren't forced to relive past trauma through shared content in our newsfeeds, we not only decrease suffering, but we also increase the potential for thriving.

In planning a path forward, it's paramount to center those who've been harmed. As Dr. Judith Herman told me, helping someone recover begins by asking most simply: "What are you feeling? What is your story? What have you experienced and what would make things right for you?"

In 2019 technologist Sydette Harry gave a keynote presentation at the Code for America Summit, telling her audience that "you only code as well as you listen." To write her speech, Harry said she pinned a survey to her Twitter profile asking six questions on technology and community. She learned from participants that "they wanted to be heard, really heard, by designers and security teams." She concluded with advice for the room, telling them not to build another app. Instead, she said, "Have another conversation. Make another conversation possible."

• • •

Most of the feedback I received during my decade at *USA Today* came in the form of harassment. Occasionally, a reader would send a thankful note. I cherished those.

Here's one I'm sharing with the sender's permission:

> Dear Alia,
>
> I am writing because I just saw your piece in the paper.
>
> I am one of 60 women who has come forward about Bill Cosby. I came forward 11 years ago as a Jane Doe willing to testify on behalf of Andrea Constand, again two years ago using my first name only, and then by my full name in July of 2015, after Mr. Cosby's own words from his deposition were revealed by the media.
>
> Women who do come forward about sexual assault undergo a process, much like the process of "coming out" in the LGBT community. We are met with disbelief, scorn and character assassination. Why would we seek fame about this traumatic experience from our lives?
>
> Thank you for your article. By highlighting the conversation we keep it in the conscious awareness of the American public. And thank you for mentioning the 60 of us who were inspired by the courage of the original 13 Jane Does who came forward in 2005. The examples of the Cosby case and the cover up by the Catholic Church were perfect.
>
> With appreciation and gratitude,
> Patricia Leary Steuer

I was reticent to share this, which is interesting, given that for the bulk of this book I mostly recounted my abuse. I think maybe I worried including the praise would read as defensive, like I was trying to prove that at least some of what I did was worth something. I ultimately decided to include it because I want people to know it can be very difficult to keep going, and we should never underestimate how meaningful it is for someone to hear that what they are doing matters. Messages like these sustained me. I can't know what will sustain you. I have offered strategies you may use to mitigate psychological harm. These actions are meaningful. They make life survivable and work possible. But they are limited. Sometimes they put us at odds with what is good for ourselves and what is right for the world.

I hope you will close this book with the knowledge that your experiences matter, that your pain is a story worth telling, and that another person's pain is a story worth listening to. I have reached the conclusion, but I hope you will view this book as a beginning, an opening, a permission to talk about things that some people would rather keep quiet.

Scholar Jacqueline Rose writes in *On Violence and On Violence Against Women* that "violence is not a subject about which anyone can believe, other than in a state of delusion, that everything has been said and done." It is the core argument of her book, Rose writes, that "violence will not diminish, let alone cease, if violence continues to be something which people turn away from, blot from their minds, prefer—at least as far as they personally are concerned—not to talk or think about."

I have thought a lot about violence against women online and have tried to share what I've learned. Now these accounts belong to you. Carry the parts you find useful. This book is a small piece of a much bigger and ever-evolving story. I hope you will keep this conversation going, that you will deepen it and address all the complexities that escaped me.

In the meantime, I'll leave you with some things I hope you'll remember.

Remember that words matter. They can be used to dominate, and they can be used to liberate. Words can be used to erase a person, a people. They can be used to make someone feel seen.

Remember that you have every right to live and work in ways

that make people wiser and more reflective. Remember that you do not have to accept things that are unacceptable.

All of us feel something, even if it's numb. Do not ignore yourself, what you need. Investigate yourself and investigate what is happening to you. Take moments to celebrate how much you have already survived.

Fight despair. Remember that you can work with others to fight despair. Find someone who will listen to your hurt, so you don't starve the world of your voice. Remember to consider what we could achieve if we truly worked together—across race, across gender, across geography, across language, across culture, across time.

Find moments to walk away from the screen. Step away from the technology we created long enough to feel gratitude for a world we were gifted.

Disabuse yourself of the notion that you do not belong.

Write down what you're thankful for. Speak your gratitude aloud, so you can hear it in your own ears.

Do not judge yourself for the sweat on your back, for the confusion in your belly, for the doubt in your throat. Do not judge the things you do to hide, the ways you hide. Do not judge your ambivalence. Do not judge your responses. Do not judge when you have to rest or even walk away. As Dr. Thema Bryant told me, "When we judge our reaction, we double our difficulty."

After Trump's 2016 win, when I interviewed scholar Kimberlé Crenshaw for a *USA Today* story on the state of gender equality, she told me not to grow inured to a culture of violence. She told me to continue to be shocked by things, to be disappointed.

"That's got to be what holds people to a commitment to insist that women's lives matter, even when it seems like they don't," she said.

I hope you will remember that the most powerful words aren't the ones that tell us our voices are too shrill, too aggressive, too brazen, too loud, that accuse us of being too sensitive or too fragile, of complaining too much.

The most powerful language belongs to us. It is the words we write and speak and shout, the words birthed through pain and rage, that nourish connection and ambition, and that fuel a magnificent resolve, the words that reject silence, that thirst for something different, that insist, to ourselves and to others, that we can defeat a suffocating violence. Our voices lead us toward our most splendid hopes.

I hear my own hopes in the voices of my children, when they speak directly, when they ask questions, when they tell me they have spotted a lie.

Last summer my oldest daughter, who lay beside me in bed when those earliest messages of hate interrupted our peace, came to me and told me words a boy used to cut her down. I asked her what she was doing when he spoke them. I could see in her face a careful contemplation.

She concluded: "I was being myself."

In my mind, I began to delicately craft a response. I did not want to alarm her, but I wanted to impress upon her early that by the time she is grown, she will have accumulated a ledger of indignities. I wanted to tell her then so that she would not be stunned by future cruelties, so she could become a vigilant and committed custodian of her self.

I was deep in my resolve, arranging my best developmentally appropriate sentences, but before I could speak, she moved on to the next point on her agenda:

"When can I get a phone?"

TOOL KIT

I NEVER WANTED to write a prescriptive book. I would never be comfortable telling another woman what she should do to mitigate her own suffering, because I don't live in her suffering. I did want to explore, without judgment, how women have managed their experiences of violence online. I realize that to do that, I have waded into many other social, cultural, and political issues, and some people have likely picked up this book only to learn something tangible that they could apply to their own lives. So I considered that perhaps it wasn't a bad idea to be able to turn to the back of this book and find some of its most practical reporting on coping neatly distilled.

Only you can define what strategies you have the capacity to engage in, what your strengths and skills look like, what your goals are, and ultimately what you want to make of your life. What I offer here are ways of coping that showed up in my interviews. I hope arranging them this way is helpful to you.

Emotion Regulation

Some amount of emotion regulation, "the processes by which individuals influence which emotions they have, when they have them, and how they experience and express these emotions," is necessary to survive daily life. But the labor of regulating emotions around oppression can also be a form of injustice. When trying to move yourself out of a difficult emotional state, consider your goals. Do you need to stay in anger to animate you? Do you need to move to a different emotion to stay safe?

Sometimes the emotions we feel during or after a difficult event may result from unhealthy thinking. If you are struggling specifically with online abuse, thinking that you did something to deserve it, that maybe you are not well equipped to handle it, consider this thought from psychotherapist Seth Gillihan:

"Maybe you could ask yourself something like 'Do any of the people I know who have expressed who they truly are in spite of society's criticism or hatred, who have really changed society, have they done it without some level of abuse? Is it worth what it's going to cost?' And the answer might be no. But someone might realize, 'Oh, nothing says this should be easy.' And there can be real relief in that. And realizing this is hard. Yeah, it's hard. Exactly. That's exactly how it is."

Humor

I have always used humor to cope with difficulty and was unsurprised to learn that laughter is beneficial for emotional and physical health. For many women humor can be its own form of meaning-making. As philosopher Cynthia Willett and historian Julie Willett write in their book *Uproarious,* humor "can serve as a source of empowerment, a strategy for outrage and truth telling, a counter to fear, a source of joy and friendship, a cathartic treatment against unmerited shame, and even a means of empathetic connection and alliance."

Anger

Philosopher Amia Srinivasan writes that anger involves "a *moral violation:* not just a violation of how one *wishes* things were, but a violation of how things *ought* to be." Anger is legitimate and necessary. The problem is not our anger but the mismanagement of our anger, which can make us sick and less effective at accomplishing our goals. As Soraya Chemaly writes in *Rage Becomes Her*, women need to develop "anger competence," a way to own and harness anger that involves "awareness, talking, listening, and strategizing."

Logging Off

Sometimes we need to log off to remind ourselves there is more to life than what is on the screen. When you cannot engage with your abuse and remain emotionally safe, consider asking people you trust to monitor your accounts. They could be friends, family members, or co-workers. "Logging off" should not be confused with "just log off," a minimizing response to women's abuse.

Counterspeech

Some women expose their abuse online to raise awareness, reject discrimination, and call for empathy. In many cases, their screenshots and commentary qualify as counterspeech, which seeks to mitigate the harms of "bad" speech with more "good" speech. While exposing abuse can risk spreading toxic content further and increasing its salience, online counterspeech can also reach many more people than offline speech, and it can warn people about the kind of bigotry they may encounter and offer ways to reject it. PEN America's message on counterspeech is that only the person being harassed will know what feels most helpful and empowering to them. For some, it's not speaking out at all. For others, it's speaking out about what is happening to them, how they're doing, or their work.

The only form of counterspeech that PEN America actively advises against is responding to hate, harassment, or abuse with more hate, harassment, and abuse. PEN America strongly advises against a direct retweet or repost of abusive or hateful content because it runs the risk of inadvertently promoting that content algorithmically. (The more engagement an abusive or hateful piece of content gets, the wider its reach could be.) Screenshotting abusive content, putting it in quotes, or paraphrasing is a better option if you want people to see what you're facing.

Compartmentalization

Many women compartmentalize to deal with the stress of online abuse. Compartmentalization is defined by the American Psychological Association as a state where "thoughts and feelings that seem to conflict or to be incompatible are isolated from each other in separate and apparently impermeable psychic compartments." When we compartmentalize, we tuck something difficult away, set it aside, distract ourselves. Compartmentalization can be either unconscious or conscious. It is a useful response when we are dealing with something that is emotionally or physiologically challenging, but psychologists say it shouldn't be a person's only response. It's important to return to whatever you put away, to unpack it, to figure out what it made you feel and how you can give yourself what you need.

Numbing and Dissociation

To protect ourselves during an overwhelming online attack, we may go numb or dissociate. Emotional numbing and dissociation can overlap. When we are emotionally numb, we are less interested in the world, in other people, and even in ourselves. Dissociation involves a breakdown in functioning during a truly overwhelming experience. It is an adaptation to trauma. It becomes a symptom when it persists in the aftermath of a trauma.

Desensitization

Many women I interviewed told me they were "desensitized" to online abuse, and some noted they felt that that was why they'd been able to remain online. While it may be useful for a woman to be less emotionally responsive to being called a slur, that diminished responsiveness can normalize violence and likely also exacts other psychological costs. Just because someone becomes desensitized to something doesn't mean that they are not distressed. Maybe you're desensitized to your abuse, but that doesn't mean you aren't experiencing other forms of disillusionment and withdrawal.

Self-Care

Meaningful self-care allows us to connect with ourselves so deeply that we have the desire and the capacity to care for something more than ourselves. Scholar Jessie Daniels told me that the way many people think about self-care is so individualized, it distracts from the kind of connections that are truly healing. In her book *Homecoming,* Dr. Thema Bryant invites us to consider: "What are you doing to cultivate your gifts? What are you doing to feed your mind and your spirit?" Self-care is self-defined. It can be exercising or journaling or dancing or scrapbooking or doing handstands. It can be spending time with people we love. Sometimes it's spending quality time with ourselves.

Purpose

Almost all the women I interviewed for this book hold a strong sense of purpose. When they were suffering, they took comfort in their purpose, used it to affirm that their voice and their work mattered. Sometimes they used it to reorient themselves. As Austrian psychiatrist and psychotherapist Viktor Frankl wrote in *Man's Search for Meaning,* "There is nothing in the world, I venture to say, that would so effectively help one to survive even the worst conditions as the knowledge that there is a meaning in one's life. There is much wisdom in the words of Nietzsche: 'He who has a *why* to live for can bear almost any *how.*'"

Culture

Culture can be a powerful protective factor for victims of violence. Marissa Indoe, who is Anishinaabe, turned to her Indigenous culture to cope with her experiences of abuse online. She leaned into Indigenous teachings, engaged in specific ceremonies and rituals, embraced mindfulness, and turned her pain into Indigenous art. She viewed her platform as a place to live her Indigenous values, and often responded to racist comments with patience and education. As Indigenous scientist Robin Wall Kimmerer writes in *Braiding Sweetgrass*, "In order for the whole to flourish, each of us has to be strong in who we are and carry our gifts with conviction." Our gifts "are not meant for us to keep. Their life is in their movement, the inhale and the exhale of our shared breath. Our work and our joy is to pass along the gift and to trust that what we put out into the universe will always come back."

Therapy

I wrote a chapter on therapy because I believed that the advice to "talk to a mental health professional" was too vague. In *Radical Feminist Therapy*, psychotherapist Bonnie Burstow emphasizes that women need a therapist who understands that "violence is absolutely integral to our experience as women." Many psychologists told me that endlessly intervening upon the individual will never bring about healing if the social, political, and economic factors contextualizing a person's suffering are not also examined. Dr. Thema Bryant advises women to look for a therapist who will help them develop a "resistance strategy."

Confrontation

Many women fight back against online abuse, making it a strategy worthy of attention, even amid concerns from online safety experts. Some women told me they experienced practical and psychological benefits when they engaged in confrontational coping. Telling a woman to never respond to her abuse may ignore women's unique vulnerabilities, their culturally specific ways of coping, and their relationship to likability and respectability politics. When a woman is considering whether and how to fight back, PEN America suggests that she ask herself if she is emotionally or psychologically ready for a potential confrontation or escalation of abuse, which may happen. She might ask herself, "Would saying something feel empowering? Do I feel it's worth it to speak directly to the harasser, given the possibility of confrontation or escalation, or can I speak out more broadly about my experiences or my work? How can I use my voice in a way that feels safe to me?"

Social Support and Solidarity

Almost every expert I interviewed for this book agreed that externalizing painful experiences of abuse can be among the most beneficial coping strategies. Feeling that we are loved and valued by other people is a key component of psychological and physical health. Social support also connects to the concept of solidarity, when we commit to working with others in a larger transformative struggle. Solidarity is how we advance collective power. For more on building support networks, see PEN America's report "The Power of Peer Support."

Safety

It's important to know what information is available about you online and decide what you're not comfortable having in the public domain. This can include your address, photos of your children, or other pieces of personal information that can be weaponized against you, including old social media posts and tweets. Consider signing up for data removal sites to have your address removed from the public domain. Check the privacy settings of all your social media accounts. Review your online profile regularly. Make sure to use two-factor authentication for all your accounts and create complex passwords that you store in a password manager. If you're attacked, consider turning all your social media accounts private. Document your abuse, and lean on others to help when the process overwhelms.

Additional Resources

Many individuals and groups are working to combat online abuse by offering direct support to victims and educating the public. The following resources provide practical advice for protecting yourself ahead of an online attack, safeguarding yourself during an attack, and working to care for yourself emotionally while navigating online abuse.

"Online Violence Response Hub," Coalition Against Online Violence, n.d., onlineviolenceresponsehub.org.

"Digital Safety: Protecting Against Online Harassment," Committee to Protect Journalists, March 20, 2023, cpj.org/2018/11/digital-safety-protecting-against-online-harassment.

"Digital Safety: Remove Personal Data from the Internet," Committee to Protect Journalists, September 4, 2019, cpj.org/2019/09/digital-safety-remove-personal-data-internet.

Autumn Slaughter and Elana Newman, "Journalists and Online Harassment," Dart Center for Journalism and Trauma, January 14, 2020, dartcenter.org/resources/journalists-and-online-harassment.

Whitney Phillips, "The Oxygen of Amplification: Better Practices for Reporting on Extremists, Antagonists, and Manipulators Online," pt. 1, Data and Society (2012), datasociety.net/pubs/oh/1_PART_1_Oxygen_of_Amplification_DS.pdf.

"Games Hotline Digital Safety Guide," Games and Online Harassment Hotline, September 10, 2023, onlinesafety.feministfrequency.com/en.

"Five Lessons for Reporting in an Age of Disinformation," *Medium,* December 28,

2018, medium.com/1st-draft/5-lessons-for-reporting-in-an-age-of-disinformation-9d98f0441722.

"Interventions to End Online Violence Against Women in Politics," National Democratic Institute, October 27, 2022, www.ndi.org/publications/interventions-end-online-violence-against-women-politics.

"How to Dox Yourself on the Internet," *New York Times,* February 27, 2020, open.nytimes.com/how-to-dox-yourself-on-the-internet-d2892b4c5954.

"Documenting Online Harassment," *Online Harassment Field Manual,* PEN America, n.d., onlineharassmentfieldmanual.pen.org/documenting-online-harassment.

"Legal Considerations," *Online Harassment Field Manual,* PEN America, n.d., onlineharassmentfieldmanual.pen.org/legal-considerations.

"Online Abuse Defense Training Program," PEN America, ongoing, pen.org/online-abuse-defense-training-program.

Online Harassment Field Manual, PEN America, n.d., onlineharassmentfieldmanual.pen.org.

"Bystander Intervention Training," Right to Be, righttobe.org.

Mental Health Guides

A Mental Health Guide for Journalists Facing Online Violence, International Women's Media Foundation, November 2022, www.iwmf.org/wp-content/uploads/2022/12/Final_IWMF-Mental-health-guide.pdf.

"Self-Care," *Online Harassment Field Manual,* PEN America, onlineharassmentfieldmanual.pen.org/self-care.

Books

I read many books that informed my reporting. Some deal directly with the Internet and online abuse, others with language, culture, emotion, and healing. I'm including some of my reading here in the hopes that it may be useful to you, too.

Complaint! by Sara Ahmed
Living a Feminist Life, by Sara Ahmed
How to Stay Safe Online, by Seyi Akiwowo
Misogynoir Transformed: Black Women's Digital Resistance, by Moya Bailey
How Emotions Are Made: The Secret Life of the Brain, by Lisa Feldman Barrett
Seven and a Half Lessons About the Brain, by Lisa Feldman Barrett
White Feminism: From the Suffragettes to the Influencers and Who They Leave Behind, by Koa Beck

Homecoming: Overcome Fear and Trauma to Reclaim Your Whole, Authentic Self, by Thema Bryant
Indigenous Peoples Rise Up: The Global Ascendency of Social Media Activism, by Bronwyn Carlson and Jeff Berglund
Brotopia: Breaking Up the Boys' Club of Silicon Valley, by Emily Chang
Rage Becomes Her: The Power of Women's Anger, by Soraya Chemaly
Hate Crimes in Cyberspace, by Danielle Citron
Cyber Racism: White Supremacy Online and the New Attack on Civil Rights, by Jessie Daniels
Broad Band: The Untold Story of the Women Who Made the Internet, by Claire L. Evans
Man's Search for Meaning, by Viktor E. Frankl
Pedagogy of the Oppressed, by Paulo Freire
The Politics of Reality: Essays in Feminist Theory, by Marilyn Frye
Nobody's Victim: Fighting Psychos, Stalkers, Pervs, and Trolls, by Carrie Goldberg
Trauma and Recovery: The Aftermath of Violence—from Domestic Abuse to Political Terror, by Judith L. Herman
Truth and Repair: How Trauma Survivors Envision Justice, by Judith L. Herman
#HashtagActivism: Networks of Race and Gender Justice, by Sarah J. Jackson, Moya Bailey, and Brooke Foucault Welles
How to Be a Woman Online, by Nina Jankowicz
Hood Feminism: Notes from the Women That a Movement Forgot, by Mikki Kendall
A Burst of Light, by Audre Lorde
Sister Outsider, by Audre Lorde
Down Girl: The Logic of Misogyny, by Kate Manne
Words That Wound: Critical Race Theory, Assaultive Speech, and the First Amendment, by Mari J. Matsuda, Charles R. Lawrence III, Richard Delgado, and Kimberlé W. Crenshaw
Words Matter, by Sally McConnell-Ginet
Reset, by Ellen Pao
This Is Why We Can't Have Nice Things, by Whitney Phillips
Check It While I Wreck It: Black Womanhood, Hip-Hop Culture, and the Public Sphere, by Gwendolyn D. Pough
Healing Justice: Holistic Self-Care for Change Makers, by Loretta Pyles
Crash Override: How Gamergate Destroyed My Life, and How We Can Win the Fight Against Online Hate, by Zoë Quinn
Behind the Screen: Content Moderation in the Shadows of Social Media, by Sarah T. Roberts
As We Have Always Done: Indigenous Freedom through Radical Resistance, by Leanne Betasamosake Simpson

ACKNOWLEDGMENTS

THANK YOU TO every woman who trusted me with their story, and to every source who shared their expertise.

Thank you to my agent, Laurie Liss, for pushing for years. Thank you to everyone at Crown who touched this project, and most importantly to my editor, Amy Li, who fully understood what I was trying to do and believed that it mattered. Thank you to the many patient thinkers who helped me refine my arguments, especially Bridget Todd, Lynne Tirrell, Sherry Hamby, Negar Fani, José Soto, Juliet Williams, Lillian Comas-Díaz, Soraya Chemaly, Taima Moeke-Pickering, Koa Beck, Brooklyne Gipson, and Kat Tenbarge. Thank you to a host of folks at the International Women's Media Foundation and PEN America for their research and insights.

Thank you to Tanya Leet, whose small act of belief and kindness may have made all the difference.

Thank you to Thea Smith for her rigorous fact-checking.

Thank you to my *USA Today* family for encouragement and support in work and in life: Patty Michalski, Anne Godlasky, Emily Brown, Cara Richardson, and Mary Nahorniak. To Anne, thank you for making me believe I could write more than a news story.

Thank you to the many patient editors over the years who let me experiment and tolerated my word count: Chrissie Thompson, Jennifer Portman, Leora Arnowitz, and Laura Trujillo. Thank you to Eve Chen for always sending the kindest notes during the worst harassment, and to Nathan Bomey for being the first person to believe I could write this book.

Thank you to Cris Beam, Jasmin Sandelson, Katie McKay, Hannah Meyer, Kimberly Rogers, Clement Yue, Lily Herold, and especially Zoe Patterson, for helping me find the courage to claim authority.

Thank you to MaryBeth Yerdon for pushing me to write full-time and for always picking up the phone. Thank you to Emily Oster and Danielle Daloia for showing me what friendship between women should look like.

Thank you to my mother-in-law, who supported my vision and offered practical help, and to Karen Dukess for the crucial introduction. Thank you to my parents, who paid me to read and pushed me to write, and for being present with my kids when I could not be.

Thank you to my husband, who never let me give up. Who never gives up.

I would have written a better book if I wasn't a mother, but I probably wouldn't have written this book if I wasn't a mother. Thank you to my daughters, who had much less of me over the past few years than they would have liked but handled my absences with patience and generosity. I hope eventually they'll read this book and feel motivated to also work toward the hope that one day, someday, all of us will move through this world and truly be free.

NOTES

Introduction

4 **"a comprehensive definition":** Sherry Hamby, "On Defining Violence, and Why It Matters," *Psychology of Violence* 7, no. 2 (2017): 167–80, doi.org/10.1037/vio0000117.

5 **Women of the blogging era:** Jill Filipovic, "Blogging While Female: How Internet Misogyny Parallels 'Real-World' Harassment," *Yale Journal of Law and Feminism* 19 (2007): 10, www.semanticscholar.org/paper/Blogging-While-Female%3A-How-Internet-Misogyny-Filipovic/c4772256643582a35ba27c64b7a834c2732dae1f.

5 **heard very clearly but did not care:** Bridget Todd, "How Black Women Tried to Save Twitter," *There Are No Girls on the Internet,* July 14, 2020, omny.fm/shows/there-are-no-girls-on-the-internet/how-black-women-tried-to-save-twitter.

5 **who are disproportionately targeted for online abuse:** Troll Patrol Project, Amnesty International and Element AI, December 2018, www.amnesty.org/en/latest/press-release/2018/12/crowdsourced-twitter-study-reveals-shocking-scale-of-online-abuse-against-women.

5 **to call attention to online harassment:** Gabriela de Oliveira and Dr. Julia Slupska, "Digital Misogynoir Report: Ending the Dehumanising of Black Women on Social Media," *Glitch* (2023), glitchcharity.co.uk/wp-content/uploads/2023/07/Glitch-Misogynoir-Report_Final_18Jul_v5_Single-Pages.pdf.

5 **continue to be erased:** Sydette Harry, "Listening to Black Women: The Innovation Tech Can't Figure Out," *Wired,* January 11, 2021, www.wired.com/story/listening-to-black-women-the-innovation-tech-cant-figure-out.

6 **White supremacists were merciless:** Eric Randall, "Trayvon Martin's Emails Leaked," *The Atlantic,* March 29, 2012, www.theatlantic.com/national/archive/2012/03/trayvon-martins-emails-leaked/329940.

6 **photos re-creating Martin's dead body:** Ryan Broderick, "'Trayvoning' Is a New Horrible Trend Where Teenagers Reenact Trayvon Martin's Death Photo," *Buzzfeed,* July 16, 2013, www.buzzfeednews.com/article/ryanhatesthis/trayvoning-is-a-new-horrible-trend-where-teenagers-reenact-t.

6 **#YourSlipIsShowing:** Rachelle Hampton, "The Black Feminists Who Saw the Alt-Right Threat Coming," *Slate,* April 23, 2019, slate.com/technology/2019/04/black-feminists-alt-right-twitter-gamergate.html.

6 **used by a community:** Sydette Harry, "Everyone Watches, Nobody Sees: How Black Women Disrupt Surveillance Theory," *Model View Culture,* October 6,

2014, modelviewculture.com/pieces/everyone-watches-nobody-sees-how-black-women-disrupt-surveillance-theory.

6 **people who called out the imposters:** Sydette Harry, "Ouroboros Outtakes: The Circle Was Never Unbroken," *Model View Culture,* December 8, 2014, modelviewculture.com/pieces/ouroboros-outtakes-the-circle-was-never-unbroken.

6 **media coverage of Gamergate:** David Nieborg and Maxwell Foxman, "Mainstreaming Misogyny: The Beginning of the End and the End of the Beginning in Gamergate Coverage," in *Mediating Misogyny*, ed. J. Vickery and T. Everbach (London: Palgrave Macmillan, 2018), doi.org/10.1007/978-3-319-72917-6_6.

6 **broader culture of abuse:** Dorothy Kim et al., "Race, Gender, and the Technological Turn: A Roundtable on Digitizing Revolution," *Frontiers: A Journal of Women Studies* 39, no. 1 (2018): 149–77, www.jstor.org/stable/10.5250/fronjwomestud.39.1.0149.

7 **"To be heard as complaining":** Sara Ahmed, *Complaint!* (Durham, NC: Duke University Press, 2021), 1.

7 **she wanted to kill herself:** Morgan Radford, "Journalists Face Online Harassment," MSNBC, April 1, 2022, www.msnbc.com/mtp-daily/watch/female-journalists-face-gender-based-online-harassment-136783429972.

8 **her essay "Oppression":** Marilyn Frye, *The Politics of Reality: Essays in Feminist Theory* (Berkeley, CA: Crossing Press, 1983), 1–16.

8 **"One of the most characteristic":** Ibid., 260.

9 **Americans reporting severe harassment increased:** Emily A. Vogels, "The State of Online Harassment," Pew Research Center, January 13, 2021, www.pewresearch.org/internet/2021/01/13/the-state-of-online-harassment.

9 **61 percent of women:** Ibid.

9 **sexually harassed or stalked:** Ibid.

9 **About half or more Black:** Ibid.

9 **people who identify as LGBTQ:** "Social Media Safety Index," GLAAD, June 15, 2023, assets.glaad.org/m/7adb1180448da194/original/Social-Media-Safety-Index-2023.pdf.

10 **"in the service of nothing":** Whitney Phillips, *This Is Why We Can't Have Nice Things: Mapping the Relationship Between Online Trolling and Mainstream Culture* (Cambridge, MA: MIT Press, 2015), 21.

10 **"linked to structural systems":** Alice Marwick, "Morally Motivated Networked Harassment as Normative Reinforcement," *Social Media + Society* 7, no. 2 (2021), doi.org/10.1177/20563051211021378.

11 **"The hate . . . I endure":** Leigh Finke, "I Am Most at Home in My Womanhood When You Hate Me for It," *Queer and Forever Here,* April 6, 2023, queerandforeverhere.blog/2023/04/06/i-am-most-my-at-home-in-my-womanhood-when-you-hate-me-for-it.

13 **"Ours is a world":** John Perry Barlow, "A Declaration of the Independence of Cyberspace," Electronic Frontier Foundation, February 8, 1996, www.eff.org/cyberspace-independence.

13 **utopian vision:** Charles Dunlop and Rob Kling, *Computerization and Controversy: Value Conflicts and Social Choices* (New York: Academic Press, 1991).

13 **"there is no race":** "MCI TV Ad 1997," YouTube, October 3, 2010, www.youtube.com/watch?v=ioVMoeCbrig.

13 **"in search of some disembodied":** Jessie Daniels, *Cyber Racism: White Supremacy Online and the New Attack on Civil Rights* (Lanham, MD: Rowman & Littlefield, 2009), 334.

14 **envisioned by men:** Emily Chang, *Brotopia: Breaking Up the Boys' Club of Silicon Valley* (New York: Portfolio/Penguin, 2018).

14 **evolve with their perspectives:** "Diversity in High Tech," a special report from the U.S. Equal Employment Opportunity Commission, n.d., www.eeoc.gov/special-report/diversity-high-tech.

14 **Discrimination is baked:** Jessie Daniels, "Race and Racism in Internet Studies: A Review and Critique," *New Media and Society* 15, no. 5 (2013): 695–719, doi.org/10.1177/1461444812462849.

14 **primary objective is profit:** Safyia Noble, *Algorithms of Oppression: How Search Engines Reinforce Racism* (New York: New York University Press, 2018), 57.

14 **worker is typically paid wages:** Sarah T. Roberts, *Behind the Screen: Content Moderation in the Shadows of Social Media* (New Haven, CT: Yale University Press, 2019), 201.

14 **a cyber civil rights legal agenda:** Danielle Keats Citron, *Hate Crimes in Cyberspace* (Cambridge, MA: Harvard University Press, 2014).

14 **made gains in achieving legal accountability:** Carrie Goldberg, *Nobody's Victim: Fighting Psychos, Stalkers, Pervs, and Trolls* (New York: Plume, 2019).

15 **staff at Google offices:** Dave Lee, "Google Staff Walk Out Over Women's Treatment," BBC, November 1, 2018, www.bbc.com/news/technology-46054202.

15 **Facebook staffers staged a virtual walkout:** Sheera Frenkel, Mike Isaac, Cecilia Kang, and Gabriel J. X. Dance, "Facebook Employees Stage Virtual Walkout to Protest Trump Posts," *New York Times,* October 10, 2021, www.nytimes.com/2020/06/01/technology/facebook-employee-protest-trump.html.

15 **deprioritized through staff reductions:** J. J. McCorvey, "Tech Layoffs Shrink 'Trust and Safety' Teams, Raising Fears of Backsliding Efforts to Curb Online Abuse," NBC News, February 10, 2023, www.nbcnews.com/tech/tech-news/tech-layoffs-hit-trust-safety-teams-raising-fears-backsliding-efforts-rcna69111.

15 **Elon Musk bought Twitter:** Jessica Guynn, "'The Bird Is Freed' as Elon Musk Now Owns Twitter. What's Next for the Social Media Giant?," *USA Today,* October 27, 2022, www.usatoday.com/story/tech/2022/10/27/elon-musk-owns-twitter-now-what/10597038002.

15 **rebranded it X:** Ryan Mac and Tiffany Hsu, "From Twitter to X: Elon Musk Begins Erasing an Iconic Internet Brand," *New York Times,* July 24, 2023, www.nytimes.com/2023/07/24/technology/twitter-x-elon-musk.html.

15 **reinstated many accounts:** Taylor Lorenz, "'Opening the Gates of Hell': Musk Says He Will Revive Banned Accounts," *Washington Post,* November 24, 2022, www.washingtonpost.com/technology/2022/11/24/twitter-musk-reverses-suspensions.

15 **dissolved its Trust and Safety Council:** Matt O'Brien and Barbara Ortutay, "Musk's Twitter Disbands Its Trust and Safety Advisory Group," Associated

Press, December 13, 2022, apnews.com/article/elon-musk-twitter-inc-technology-business-a9b795e8050de12319b82b5dd7118cd7.

15 **promote conspiracy theories:** Marshall Cohen, "Elon Musk Peddles Debunked 2020 Election Conspiracies at First Solo Town Hall Supporting Trump," CNN, October 17, 2024, www.cnn.com/2024/10/17/media/elon-musk-dominion-voting-misinformation/index.html.

15 **a rise in authoritarianism:** Sarah Repucci and Amy Slipowitz, "Freedom in the World 2022: The Global Expansion of Authoritarian Rule," Freedom House, February 2022, freedomhouse.org/sites/default/files/2022-02/FIW_2022_PDF_Booklet_Digital_Final_Web.pdf.

15 **pandemic put us online:** Colleen McClain et al., "The Internet and the Pandemic," Pew Research Center, September 1, 2021, www.pewresearch.org/internet/2021/09/01/the-internet-and-the-pandemic.

15 **AI is threatening to pollute:** Brianna Scott, Jeanette Woods, and Alisa Chang, "How AI Could Perpetuate Racism, Sexism and Other Biases in Society," NPR, July 19, 2023, www.npr.org/2023/07/19/1188739764/how-ai-could-perpetuate-racism-sexism-and-other-biases-in-society.

15 **Disinformation is damaging:** "Gendered Disinformation: Tactics, Themes, and Trends by Foreign Malign Actors," U.S. Department of State, March 27, 2023, www.state.gov/gendered-disinformation-tactics-themes-and-trends-by-foreign-malign-actors.

15 **Internet subcultures are manipulating the media:** Alice Marwick and Rebecca Lewis, "Media Manipulation and Disinformation Online," Data and Society Research Institute, May 15, 2017, datasociety.net/wp-content/uploads/2017/05/DataAndSociety_MediaManipulationAndDisinformationOnline-1.pdf.

15 **monetization of misogyny:** Lucina Di Meco, "Monetizing Misogyny: Gendered Disinformation and the Undermining of Women's Rights and Democracy Globally," #ShePersisted, February 2023, she-persisted.org/wp-content/uploads/2023/02/ShePersisted_MonetizingMisogyny.pdf.

15 **danger to free speech:** "ACLU Responds to Election of Donald Trump," ACLU, www.aclu.org/press-releases/aclu-responds-to-election-of-donald-trump, accessed on November 6, 2024.

16 **poured into Washington, D.C.:** Christina Barron, "Hundreds of Thousands Come to Washington for Women's March," *Washington Post,* January 21, 2017, www.washingtonpost.com/lifestyle/kidspost/hundreds-of-thousands-come-to-washington-for-womens-march/2017/01/21/97daf72a-dff3-11e6-ad42-f3375f271c9c_story.html.

16 **home on the Upper East Side:** "A Home for a Movement," Google Arts and Culture, artsandculture.google.com/story/a-home-for-a-movement/jgIi1197UxiDIA.

16 **famous black-and-white photo:** Dan Wynn, "Gloria Steinem and Dorothy Pitman Hughes" (photo), 1971, National Portrait Gallery, npg.si.edu/object/npg_NPG.2005.121.

16 **her friend Wilma Mankiller:** "Putting History into Action: Gloria and Wilma School Links Activists and Archives," Smith College, July 20, 2015, www.smith.edu/news-events/news/putting-history-action-gloria-wilma-school-links-activists-and-archives-0.

17 **I wrote for Women's History Month:** Alia E. Dastagir, "What Do Men Get That Women Don't? Here Are a Few Things," *USA Today,* March 1, 2017, www.usatoday.com/story/news/2017/03/01/2017-womens-history-month/98247518.

18 **Zoë Quinn wrote in their memoir:** Zoë Quinn, *Crash Override: How Gamergate (Nearly) Destroyed My Life, and How We Can Win the Fight Against Online Hate* (New York: PublicAffairs, 2017), 194.

18 **can reduce the experience of the pain:** Adriaan Louw, Emilio Louie, J. Puentedura, and Kory Zimney, "Teaching Patients About Pain: It Works, But What Should We Call It?," *Physiotherapy Theory and Practice* 32, no. 5 (2016): 328–31, doi.org/10.1080/09593985.2016.1194669.

19 **Sarachild believed that knowing:** Kathie Sarachild, "Consciousness-Raising: A Radical Weapon," presentation at the First National Conference of Stewardesses for Women's Rights in New York City, March 12, 1973, www.rapereliefshelter.bc.ca/wp-content/uploads/2021/03/Feminist-Revolution-Consciousness-Raising-A-Radical-Weapon-Kathie-Sarachild.pdf.

Chapter 1: Just Ignore It

22 **"I don't really understand":** Alexandria Onuoha, "Dancing Around White Supremacy," *Bates Student,* March 22, 2020, thebatesstudent.com/18926/forum/dancing-around-white-supremacy-2.

22 **"Dancing Around White Supremacy":** Ibid.

23 **less than two thousand:** "Bates at a Glance: Fast Facts," Bates College, www.bates.edu/about/facts.

24 **termed *misogynoir*:** Moya Bailey, *Misogynoir Transformed: Black Women's Digital Resistance* (New York: New York University Press, 2021), xiii.

24 **"such as 'the angry Black woman'":** Gabriela de Oliveira and Dr. Julia Slupska, "Digital Misogynoir Report: Ending the Dehumanising of Black Women on Social Media," *Glitch,* 2023, glitchcharity.co.uk/wp-content/uploads/2023/07/Glitch-Misogynoir-Report_Final_18Jul_v5_Single-Pages.pdf.

24 **84 percent more likely:** Troll Patrol Project, Amnesty International and Element AI, December 2018, www.amnesty.org/en/latest/press-release/2018/12/crowdsourced-twitter-study-reveals-shocking-scale-of-online-abuse-against-women.

24 **"doubly denigrated":** Michael Halpin et al., "Men Who Hate Women: The Misogyny of Involuntarily Celibate Men," *New Media and Society* (2023), doi.org/10.1177/14614448231176777.

24 **like 4chan:** Adrienne Massanari, "#Gamergate and the Fappening: How Reddit's Algorithm, Governance, and Culture Support Toxic Technocultures," *New Media and Society* 19, no. 3 (2017): 329–46.

24 **and Reddit:** Kyle Wagner, "The Future of the Culture Wars Is Here, and It's Gamergate," *Deadspin,* October 14, 2014, deadspin.com/the-future-of-the-culture-wars-is-here-and-its-gamerga-1646145844.

24 **the text-based chat system IRC:** Zoë Quinn, *Crash Override: How Gamergate (Nearly) Destroyed My Life, and How We Can Win the Fight Against Online Hate* (New York: PublicAffairs, 2017), 61.

25 **Black women are gamers, too:** Sierra Leone Starks, "Black Girl Gamers Band Together Against 2023's Final Boss: Loneliness," *Allure,* August 22, 2023, www.allure.com/story/black-girl-gaming-groups-loneliness.

26 **women voted for Trump:** "For Most Trump Voters, 'Very Warm' Feelings for Him Endured," Pew Research Center, August 9, 2018, www.pewresearch.org/politics/2018/08/09/for-most-trump-voters-very-warm-feelings-for-him-endured.

26 **white women helped:** Emily Guskin, Chris Alcantara, and Janice Kai Chen, "Exit Polls from the 2024 Presidential Election," *Washington Post,* November 6, 2024, www.washingtonpost.com/elections/interactive/2024/exit-polls-2024-election.

26 **uptick in harassment and intimidation:** Mark Potok, "The Trump Effect," *Intelligence Report,* Southern Poverty Law Center, Spring 2017, www.splcenter.org/fighting-hate/intelligence-report/2017/trump-effect.

26 **online harassment has gotten more severe:** Emily A. Vogels, "The State of Online Harassment," Pew Research Center, January 13, 2021, www.pewresearch.org/internet/2021/01/13/the-state-of-online-harassment.

26 **have become more visible:** Michelle Ferrier, "Attacks and Harassment: The Impact on Female Journalists and Their Reporting," TrollBusters and International Women's Media Foundation, 2018, www.iwmf.org/wp-content/uploads/2018/09/Attacks-and-Harassment.pdf.

26 **a separate global study:** Julie Posetti and Nabeelah Shabbir, "The Chilling: A Global Study of Online Violence Against Women Journalists," International Center for Journalists and UNESCO, November 2, 2022, www.icfj.org/sites/default/files/2022-11/ICFJ_UNESCO_The%20Chilling_2022_1.pdf.

26 **spike in xenophobia and hate speech:** Amy Qin, "Xenophobia and Hate Speech Are Spiking Heading Into the Election," *New York Times,* November 1, 2024, www.nytimes.com/2024/11/01/us/xenophobia-hate-speech-increase-election.html.

26 **onslaught of online abuse:** Isabelle Frances-Wright and Moustafa Ayad, "'Your Body, My Choice': Hate and Harassment Towards Women Spreads Online," *Digital Dispatches,* Institute for Strategic Dialogue, November 8, 2024, www.isdglobal.org/digital_dispatches/your-body-my-choice-hate-and-harassment-towards-women-spreads-online.

27 **collaborated with PEN America:** "Advice from a Psychologist," *Online Harassment Field Manual,* PEN America, onlineharassmentfieldmanual.pen.org/advice-from-a-psychologist.

27 **our brain's most important job:** Lisa Feldman Barrett, *Seven and a Half Lessons About the Brain* (Boston: Houghton Mifflin Harcourt, 2020), 1–12.

27 **Words have potent effects:** Ibid., 89–90.

28 **"pour content into words":** Sally McConnell-Ginet, *Words Matter* (New York: Cambridge University Press, 2020), 1–2.

30 **"demure" trend:** Charmaine Patterson, "Jools Lebron, TikTok Star Behind 'Demure' Trend, Makes Late Night Debut," *People,* August 20, 2014, people.com/jools-lebron-demure-jimmy-kimmel-8697972.

30 ***fatty*:** @joolieannie, www.tiktok.com/t/ZPRKvxTLw.

30 **full glam:** @joolieannie, www.tiktok.com/t/ZPRKvfNoy.
30 **for not sweating off her makeup:** @joolieannie, www.tiktok.com/@joolieannie/video/7272506868315065643?_r=1&_t=8nCE5J3YPug.
30 **met a man:** @joolieannie, www.tiktok.com/@joolieannie/video/7272418160484125995?_r=1&_t=8nCEagmzgAo.
30 **Content creators with large:** Kurt Thomas et al., "'It's Common and a Part of Being a Content Creator': Understanding How Creators Experience and Cope with Hate and Harassment Online," in *Proceedings of the 2022 CHI Conference on Human Factors in Computing Systems (CHI '22)* (New York: Association for Computing Machinery, 2022), Article 121, 1–15, doi.org/10.1145/3491102.3501879.
32 **"surviving the relentlessness of sexism":** Sara Ahmed, *Living a Feminist Life* (Durham, NC: Duke University Press, 2017), 36.
32 **"the processes by which individuals":** James J. Gross, "The Emerging Field of Emotion Regulation: An Integrative Review," *Review of General Psychology* 2, no. 3 (1998): 271–99, doi.org/10.1037/1089-2680.2.3.271.
32 **two common strategies:** James J. Gross, "Emotion Regulation in Adulthood: Timing Is Everything," *Current Directions in Psychological Science* 10, no. 6 (2001): 214–19, doi.org/10.1111/1467-8721.00152.
32 **more likely to suppress their emotions:** Tyia K. Wilson and Amy L. Gentzler, "Emotion Regulation and Coping with Racial Stressors Among African Americans Across the Lifespan," *Developmental Review* 61 (2021), doi.org/10.1016/j.dr.2021.100967.
32 **physiologically and cognitively taxing:** Ibid.
34 **Discrimination is terrible for health:** Alia E. Dastagir, "Microaggressions Don't Just 'Hurt Your Feelings,'" *USA Today,* February 28, 2018, www.usatoday.com/story/news/2018/02/28/what-microaggressions-small-slights-serious-consequences/362754002.
34 **turn away from valid emotional responses:** Amia Srinivasan, "The Aptness of Anger," *Journal of Political Philosophy* 26, no. 2 (2018): 123–44, users.ox.ac.uk/~corp1468/Research_files/jopp.12130.pdf.
34 **"demand faced by victims of oppression":** Alfred Archer and Georgina Mills, "Anger, Affective Injustice, and Emotion Regulation," *Philosophical Topics* 47, no. 2 (2019): 75–94, www.jstor.org/stable/26948107.
35 **reappraisal is not always healthy:** Brett Q. Ford and Allison S. Troy, "Reappraisal Reconsidered: A Closer Look at the Costs of an Acclaimed Emotion-Regulation Strategy," *Current Directions in Psychological Science* 28, no. 2 (2019): 195–203, doi.org/10.1177/0963721419827526.
36 **three emotion regulation strategies:** Ajua Duker, Dorainne J. Green, Ivuoma N. Onyeador, and Jennifer A. Richeson, "Managing Emotions in the Face of Discrimination: The Differential Effects of Self-Immersion, Self-Distanced Reappraisal, and Positive Reappraisal," *Emotion* 22, no. 7 (2022): 1435–49, doi.org/10.1037/emo0001001.
36 **reappraisal is less helpful:** José A. Soto et al., "Strength in Numbers? Cognitive Reappraisal Tendencies and Psychological Functioning Among Latinos in the Context of Oppression," *Cultural Diversity and Ethnic Minority Psychology* 18, no. 4 (2012): 384–94, doi.org/10.1037/a0029781.

37 **for members of East Asian cultures:** José A. Soto et al., "Is Expressive Suppression Always Associated with Poorer Psychological Functioning? A Cross-Cultural Comparison Between European Americans and Hong Kong Chinese," *Emotion* 11, no. 6 (2011): 1450–55, doi.org/10.1037/a0023340.

37 **Global Network on Extremism and Technology:** Alexandria Onuoha, "Digital Misogynoir and White Supremacy: What Black Feminist Theory Can Teach Us About Far Right Extremism," Global Network on Extremism and Technology, August 9, 2021, gnet-research.org/2021/08/09/digital-misogynoir-and-white-supremacy-what-black-feminist-theory-can-teach-us-about-far-right-extremism.

37 **far-right misogynoir:** Alexandria C. Onuoha, Miriam R. Arbeit, and Seanna Leath, "Far-Right Misogynoir: A Critical Thematic Analysis of Black College Women's Experiences with White Male Supremacist Influences," *Psychology of Women Quarterly* 47, no. 2 (2023): 180–96, doi.org/10.1177/03616843231156872.

37 **"Anytime I write something":** Elly Belle, "Social Media Safety: How to Protect Yourself Online," *Teen Vogue,* January 28, 2022, www.teenvogue.com/story/social-media-safety-how-to-protect-yourself-online.

Chapter 2: Words Will ~~Never~~ Hurt Me

41 **It goes viral:** Brooke Rolfe, "Comedian Exposes Man Who Sent Her Unsolicited Nudes, Vulgar Messages on Valentine's Day," *New York Post,* February 16, 2023, nypost.com/2023/02/16/maria-decotis-exposes-man-who-sent-her-unsolicited-nudes.

42 **sexually harassed online:** Emily A. Vogels, "The State of Online Harassment," Pew Research Center, January 13, 2021, www.pewresearch.org/internet/2021/01/13/the-state-of-online-harassment.

42 **research on female college students:** Jennifer A. Scarduzio, Sarah E. Sheff, and Mathew Smith, "Coping and Sexual Harassment: How Victims Cope Across Multiple Settings," *Archives of Sexual Behavior* 47 (2018): 327–40, doi.org/10.1007/s10508-017-1065-7.

42 **Over half of U.S. women:** "Fast Facts: Preventing Sexual Violence," Centers for Disease Control and Prevention, www.cdc.gov/violenceprevention/sexualviolence/fastfact.html.

43 **female gamers who use:** Lavinia McLean and Mark D. Griffiths, "Female Gamers' Experience of Online Harassment and Social Support in Online Gaming: A Qualitative Study," *International Journal of Mental Health Addiction* 17 (2019): 970–94, doi.org/10.1007/s11469-018-9962-0.

43 **the sound of a woman's voice:** Jeffrey H. Kuznekoff and Lindsey M. Rose, "Communication in Multiplayer Gaming: Examining Player Responses to Gender Cues," *New Media and Society* 15, no. 4 (2013): 541–56, doi.org/10.1177/1461444812458271.

44 **are also social and political:** Sara Ahmed, *The Cultural Politics of Emotion* (London: Routledge, 2004).

44 **the politics of emotion:** Stephanie A. Shields, "The Politics of Emotion in

Everyday Life: 'Appropriate' Emotion and Claims on Identity," *Review of General Psychology* 9, no. 1 (2005): 3–15, doi.org/10.1037/1089-2680.9.1.3.

45 **"bitch sessions":** Kathie Sarachild, "Consciousness-Raising: A Radical Weapon," presentation at the First National Conference of Stewardesses for Women's Rights in New York City, March 12, 1973, www.rapereliefshelter .bc.ca/wp-content/uploads/2021/03/Feminist-Revolution-Consciousness -Raising-A-Radical-Weapon-Kathie-Sarachild.pdf.

45 **"The central feature of emotion":** Stephanie A. Shields, Heather J. MacArthur, and Kaitlin McCormick-Huhn, "The Gendering of Emotion and the Psychology of Women," in *APA Handbook of the Psychology of Women: History, Theory, and Battlegrounds,* ed. C. B. Travis et al., 189–206 (New York: American Psychological Association, 2018), doi.org/10.1037 /0000059-010.

46 **a torrent of online abuse:** Lyz Lenz, "When the Mob Comes," *Men Yell at Me,* March 31, 2021, lyz.substack.com/p/when-the-mob-comes.

46 **crude Bernie Sanders supporter:** Bridget Todd, "Disinformed: So You're the Target of an Online Hate Campaign. Now What?" *There Are No Girls on the Internet* (podcast), Episode 212, April 6, 2021, www.tangoti.com/episode -212.

46 **profile of Tucker Carlson:** Lyz Lenz, "The Mystery of Tucker Carlson," *Columbia Journalism Review,* September 5, 2018, www.cjr.org/the_profile /tucker-carlson.php.

47 **alt-right figure Richard Spencer:** Lyz Lenz, "You Should Care That Richard Spencer's Wife Says He Abused Her," *HuffPost,* January 13, 2019, www .huffpost.com/entry/richard-spencer-nina-kouprianova-divorce-abuse_n _5c2fc90ee4b0d75a9830ab69.

47 **sense of social connection:** Naomi Eisenberger, "The Pain of Social Disconnection: Examining the Shared Neural Underpinnings of Physical and Social Pain," *Nature Reviews Neuroscience* 13 (2012): 421–34, doi.org/10.1038 /nrn3231.

48 **to enforce social norms:** Alice Marwick, "Morally Motivated Networked Harassment as Normative Reinforcement," *Social Media + Society* 7, no. 2 (2021), doi.org/10.1177/20563051211021378.

48 **Osteogenesis imperfecta is a genetic:** "Osteogenesis Imperfecta: Overview of Osteogenesis Imperfecta," National Institute of Arthritis and Musculoskeletal and Skin Diseases, www.niams.nih.gov/health-topics/osteogenesis-imperfecta.

49 **lawyer Carrie Goldberg:** "Your show scrambled me, @MariaDeCotis! Left in such a chaos of sorrow horror and hilarity. Grateful to have the experience with @AnnieSeifullah who is among many many things to me, my content soul mate. Folks: Maria is a singular phenomenon & you should see her live," @cagoldberglaw, Tweet, July 14, 2023, twitter.com/cagoldberglaw/status /1679945839295299584.

50 **"I'm actually such a good comedian":** Maria DeCotis, "Emotionally Unreasonable," Union Hall, Brooklyn, July 13, 2023.

50 **"can serve as a source":** Cynthia Willett and Julie Willett, *Uproarious: How Feminists and Other Subversive Comics Speak Truth* (Minneapolis: University of Minnesota Press, 2019), 2.

50 **a buffer against stress:** Rod A. Martin and Thomas E. Ford, *The Psychology of Humor: An Integrative Approach* (Elsevier, 2018), 295, 316.

50 **physical health benefits:** Ibid., 339.

50 **fostering social connection:** Willett and Willett, *Uproarious,* 107.

51 **put up a fake article:** Shifu Careaga, "The Leftist Anti-MIMS of Bunkish Propaganda: A Study in POS Simpreme Behavior, in Allegiance to a Grander POS Supreme Power Arch Diocese of Evil," Academia.edu, November 2022, www.academia.edu/91387587/MESS0031_MIM.

51 **man plowed down several Toronto:** Alia E. Dastagir, "Incels, Alek Minassian and the Dangerous Idea of Being Owed Sex," *USA Today,* April 26, 2018, www.usatoday.com/story/news/2018/04/26/incel-rebellion-alek-minassian-sexual-entitlement-mens-rights-elliot-rodger/550635002.

51 **Eric Schneiderman was the New York:** Alia E. Dastagir, "Powerful Men Like Eric Schneiderman Show 'Hypocrisy'—But It Doesn't End with Them," *USA Today,* May 9, 2018, www.usatoday.com/story/news/nation/2018/05/09/eric-schneiderman-me-too-hypocrisy/590094002.

51 **accused of abusing women:** Jane Mayer and Ronan Farrow, "Four Women Accuse New York's Attorney General of Physical Abuse," *New Yorker,* May 7, 2018, www.newyorker.com/news/news-desk/four-women-accuse-new-yorks-attorney-general-of-physical-abuse.

52 **defines the feminist killjoy:** Sara Ahmed, "No," *feministkilljoys,* June 30, 3017, feministkilljoys.com/2017/06/30/no.

53 **create their own counterpublics:** Amy Billingsley, "Laughing Against Patriarchy: Humor, Silence, and Feminist Resistance," University of Oregon, 2013, pages.uoregon.edu/uophil/files/Philosophy_Matters_Submission_Marvin_Billingsley.pdf.

53 **London billboard in the 1970s:** Jill Posener, *Spray It Loud* (Pandora, 1982), 13.

53 **"Penis C.K.":** "Wow, looking back at the 'Louie' opening, there were definitely signs . . . ," Kelsey Caine (@kelsey_caine), Twitter, August 31, 2018, twitter.com/kelsey_caine.

53 **after five women:** Melena Ryzik, Cara Buckley, and Jodi Kantor, "Louis C.K. Is Accused by 5 Women of Sexual Misconduct," *New York Times,* November 9, 2017, www.nytimes.com/2017/11/09/arts/television/louis-ck-sexual-misconduct.html.

55 **"I genuinely think":** Maria DeCotis (@MariaDeCotis), Twitter, March 19, 2023, twitter.com/MariaDeCotis/status/1637646739539922944.

55 **Twitter removed the post:** Maria DeCotis (@MariaDeCotis), Twitter, February 16, 2023, twitter.com/MariaDeCotis/status/1626341812939145217.

55 **"We blame white male violence":** DeCotis, "Emotionally Unreasonable."

Chapter 3: Where Bodies Live

57 **a study on uterine cancer:** See Rebecca S. Dresser, "Wanted. Single, White Male for Medical Research," *Hastings Center Report* 22, no. 1 (1992): 24–29.

57 **an egregious research gap:** Dresser, "Wanted. Single, White Male."

58 **"Every gun ownership":** Dana Loesch (@DLoesch), Twitter, April 13, 2013, twitter.com/DLoesch/status/323261606349918209.

58 **maternal mortality rates:** Eugene Declercq, Ruby Barnard-Mayers, Laurie C. Zephyrin, and Kay Johnson, "The U.S. Maternal Health Divide: The Limited Maternal Health Services and Worse Outcomes of States Proposing New Abortion Restrictions" (issue brief), Commonwealth Fund, December 14, 2022, www.commonwealthfund.org/publications/issue-briefs/2022/dec/us-maternal-health-divide-limited-services-worse-outcomes.

58 **"Do dead women from forced pregnancy matter?":** Leah Torres (@LeahNTorres), Twitter, April 13, 2013, twitter.com/LeahNTorres/status/323263866261868545.

58 **She saw a tweet asking:** *Dr. Leah N. Torres vs The Western Journal, LLC; Liftable Media, Inc.; and The Daily Caller, Inc.*

59 **"You know fetuses can't scream, right?":** Ibid.

59 **falsely claiming Leah performed illegal abortions:** Ibid.

59 **filed a lawsuit:** Ibid.

59 **reached a settlement:** "Statement from the Daily Caller," *Daily Caller,* March 19, 2021, dailycaller.com/2021/03/19/statement.

59 **the outlet issued a correction:** Justin Caruso, "'Villain Explaining His Plan'—Ben Shapiro Slams Abortion Doctor for Sickening Tweet," *Daily Caller,* March 14, 2018, dailycaller.com/2018/03/14/ben-shapiro-abortion-tweet.

59 **took away the temporary medical license:** Greg Garrison, "Alabama Suspends License of Doctor for Tuscaloosa Abortion Clinic," *Birmingham News,* September 10, 2020, www.al.com/news/2020/09/alabama-suspends-license-of-doctor-for-tuscaloosa-abortion-clinic.html.

59 **"public statements related":** *Alabama State Board of Medical Examiners v. Leah N. Torres,* uploads.guim.co.uk/2023/03/10/AL_State_Board_of_Medical_Examiners_VS_Leah_N_Torres_MD.pdf.

59 **granted her Alabama medical license:** Poppy Noor, "She Was One of Alabama's Last Abortion Doctors. Then They Came for Everything She Had," *Guardian,* March 22, 2023, www.theguardian.com/world/2023/mar/22/alabama-last-abortion-doctor-leah-torres.

59 **reversed *Roe v. Wade*:** Robert Barnes and Ann E. Marimow, "Supreme Court Ruling Leaves States Free to Outlaw Abortion," *Washington Post,* June 24, 2022, www.washingtonpost.com/politics/2022/06/24/supreme-court-ruling-abortion-dobbs.

59 **passed abortion bans:** Elizabeth Nash and Isabel Guarnieri, "Six Months Post-Roe, 24 US States Have Banned Abortion or Are Likely to Do So: A Roundup" (policy analysis), Guttmacher Institute, January 2023, www.guttmacher.org/2023/01/six-months-post-roe-24-us-states-have-banned-abortion-or-are-likely-do-so-roundup.

60 **wrote a profile of Leah:** Noor, "Alabama's Last Abortion Doctors."

61 **A study on cyberstalking:** Carsten Maple, Emma Short, and Antony Brown, "Cyberstalking in the United Kingdom: Analysis of the ECHO Pilot Survey," National Centre for Cyberstalking Research, University of Bedfordshire, 2011, uobrep.openrepository.com/bitstream/handle/10547/270578/ECHO_Pilot_Final.pdf?sequence=1&isAllowed=y.

61 **A global study of online violence:** Julie Posetti and Nabeelah Shabbir, "The Chilling: A Global Study of Online Violence Against Women Journalists,"

International Center for Journalists and UNESCO, November 2, 2022, www.icfj.org/sites/default/files/2022-11/ICFJ_UNESCO_The%20Chilling_2022_1.pdf.

61 **"contribute significantly to non-linguistic actions":** Sally McConnell-Ginet, *Words Matter* (Cambridge: Cambridge University Press, 2020), 137–38.

61 **11 murders:** "2022 Violence & Disruption Statistics," National Abortion Federation, May 11, 2023, prochoice.org/wp-content/uploads/2022-VD-Report-FINAL.pdf.

62 **law enforcement lacks training:** Danielle Keats Citron, *Hate Crimes in Cyberspace* (Cambridge, MA: Harvard University Press, 2014), 20.

63 **affective injustice:** Shiloh Whitney, "Anger Gaslighting and Affective Injustice," *Philosophical Topics,* philarchive.org/rec/WHIAGA-2.

65 **"experience physiological symptoms and emotional distress":** Mari J. Matsuda, "Public Response to Racist Speech: Considering the Victim's Story," in Mari J. Matsuda et al., *Words That Wound: Critical Race Theory, Assaultive Speech, and the First Amendment* (London: Routledge, 1993), 24.

65 **raises your stress level:** "Understanding the Stress Response," Harvard Health Publishing, Harvard Medical School, July 6, 2020, www.health.harvard.edu/staying-healthy/understanding-the-stress-response.

65 **shuts down your appetite:** "Why Stress Causes People to Overeat," Harvard Health Publishing, Harvard Medical School, February 15, 2021, www.health.harvard.edu/staying-healthy/why-stress-causes-people-to-overeat.

66 **conducted her thesis:** Allison Cipriano, "Humor and Personality: An Exploration of the Predictors and Effects of Rape Humor," Ball State University, 2018.

68 **entering the age of AI:** Emily M. Bender, Timnit Gebru, Angelina McMillan-Major, and Shmargaret Shmitchell, "On the Dangers of Stochastic Parrots: Can Language Models Be Too Big?" in *Proceedings of the 2021 ACM Conference on Fairness, Accountability, and Transparency,* 610–23 (New York: Association for Computing Machinery, 2021), dl.acm.org/doi/10.1145/3442188.3445922.

68 **perpetuate gender and racial bias:** Brianna Scott, Jeanette Woods, and Alisa Chang, "How AI Could Perpetuate Racism, Sexism and Other Biases in Society," *All Things Considered,* NPR, July 19, 2023, www.npr.org/2023/07/19/1188739764/how-ai-could-perpetuate-racism-sexism-and-other-biases-in-society.

68 **"disinformation and online abuse campaigns easier":** Summer Lopez, "Speech in the Machine: Generative AI's Implications for Free Expression," PEN America, July 31, 2023, pen.org/report/speech-in-the-machine.

68 **"Brief withdrawals from your body budget":** Lisa Feldman Barrett, *Seven and a Half Lessons About the Brain* (Boston: Houghton Mifflin Harcourt, 2020), 90.

69 **Chronic stress lowers immunity:** Firdaus S. Dhabhar, "Effects of Stress on Immune Function: The Good, the Bad, and the Beautiful," *Immunologic Research* 58 (2014): 193–210, doi.org/10.1007/s12026-014-8517-0.

69 **It is linked to inflammation:** Agnese Mariotti, "The Effects of Chronic Stress on Health: New Insights Into the Molecular Mechanisms of Brain-

Body Communication," *Future Sci OA* 1, no. 3 (2015), doi.org/10.4155/fso.15.21.

69 **associated with cognitive impairment:** Ambar Kulshreshtha et al., "Association of Stress with Cognitive Function Among Older Black and White US Adults," *JAMA Network Open* 6, no. 3 (2023): e231860, jamanetwork.com/journals/jamanetworkopen/fullarticle/2802090.

69 **change the structure of our brains:** Mariotti, "Effects of Chronic Stress."

69 **offspring can be affected by trauma:** Rachel Yehuda and Amy Lehrner, "Intergenerational Transmission of Trauma Effects: Putative Role of Epigenetic Mechanisms," *World Psychiatry* 17 (2018): 243–57, doi.org/10.1002/wps.20568.

69 **Many of her patients:** Noor, "Alabama's Last Abortion Doctors."

69 **low-income backgrounds:** Alice Miranda Ollstein, "This Alabama Health Clinic Is Under Threat. It Doesn't Provide Abortions," *Politico,* May 29, 2023, www.politico.com/news/magazine/2023/05/29/alabama-abortion-clinic-problem-00096020.

70 **"weathers" marginalized bodies:** Alia E. Dastagir, "Microaggressions Don't Just 'Hurt Your Feelings,'" *USA Today,* February 28, 2018, www.usatoday.com/story/news/2018/02/28/what-microaggressions-small-slights-serious-consequences/362754002.

70 **murdered her father:** J. David Goodman and Al Baker, "Wave of Protests After Grand Jury Doesn't Indict Officer in Eric Garner Chokehold Case," *New York Times,* December 3, 2014, www.nytimes.com/2014/12/04/nyregion/grand-jury-said-to-bring-no-charges-in-staten-island-chokehold-death-of-eric-garner.html.

70 **Shafiqah Hudson:** Penelope Green, "Shafiqah Hudson, Who Fought Trolls on Social Media, Dies at 46," *New York Times,* March 5, 2024, www.nytimes.com/2024/03/05/us/shafiqah-hudson-dead.html.

70 **"feminist snap":** Sara Ahmed, *Living a Feminist Life* (Durham, NC: Duke University Press, 2017), 189.

72 **National Women's Law Center:** "Defiant: Abortion Provider Story from Dr. Leah," National Women's Law Center, May 2, 2023, www.youtube.com/watch?v=YRZX_ef-fSQ.

72 **"a *moral violation*":** Amia Srinivasan, "The Aptness of Anger," *Journal of Political Philosophy* 26, no. 2 (2018): 123–44, users.ox.ac.uk/~corp1468/Research_files/jopp.12130.pdf.

72 **the more anger a woman feels:** Kimberly Matheson and Hymie Anisman, "Anger and Shame Elicited by Discrimination: Moderating Role of Coping on Action Endorsements and Salivary Cortisol," *European Journal of Social Psychology* 39, no. 2 (2009): 163–85, doi.org/10.1002/ejsp.522.

72 **"Anger is loaded with information":** Audre Lorde, "The Uses of Anger: Women Responding to Racism," in *Sister Outsider* (Berkeley, CA: Crossing Press, 1984), 120.

72 **Black women who express anger:** Wendy Ashley, "The Angry Black Woman: The Impact of Pejorative Stereotypes on Psychotherapy with Black Women," *Social Work in Public Health* 29, no. 1 (2014): 27–34, doi.org/10.1080/19371918.2011.619449.

73 **repercussions of expressing anger:** Cathleen A. Power, Elizabeth R. Cole, and Barbara L. Fredrickson, "Poor Women and the Expression of Shame and Anger: The Price of Breaking Social Class Feeling Rules," *Feminism and Psychology* 21, no. 2 (2011): 179–97, doi.org/10.1177/0959353510384125.

73 **"is an emotion of privilege":** Stephanie A. Shields, Heather J. MacArthur, and Kaitlin McCormick-Huhn, "The Gendering of Emotion and the Psychology of Women," in *APA Handbook of the Psychology of Women: History, Theory, and Battlegrounds*, ed. C. B. Travis et al., 189–206 (New York: American Psychological Association, 2018), doi.org/10.1037/0000059-010.

73 **sometimes violence:** Grace Jacob et al., "A Systematic Review of Black People Coping with Racism: Approaches, Analysis, and Empowerment," *Perspectives on Psychological Science* 18, no. 2 (2023): 392–415, doi.org/10.1177/17456916221100509.

73 **"Victims of injustice often face":** Srinivasan, "Anger."

73 **"anger in white men":** Soraya Chemaly, *Rage Becomes Her: The Power of Women's Anger* (New York: Atria Books, 2018).

73 **Mismanaging anger can lead:** Ibid., 52–54.

74 **"anger competence":** Ibid., 261.

74 **"awareness, talking, listening, and strategizing":** Ibid., 275.

74 **self-awareness about their anger:** Ibid., 260–81.

74 **Leah gave a talk about abortion:** "Dr. Leah Torres: The Work and Witness of an Abortion Provider," All Saints Church Pasadena, YouTube, October 3, 2018, www.youtube.com/watch?v=x8WjRZhX90E.

Chapter 4: Thin Skin

76 **rise in coordinated violence:** "The Epidemic of Violence Against the Transgender and Gender Non-Conforming Community in the United States," Human Rights Campaign Foundation, November 2023, reports.hrc.org/an-epidemic-of-violence-2023.

76 **passed bill after bill:** "National State of Emergency: Know Your Rights—Summer 2023," Human Rights Campaign, August 17, 2023, www.hrc.org/resources/national-state-of-emergency-know-your-rights-summer-2023?_ga=2.138277040.1838425780.1704063378-1176275750.1704063377.

77 **created new definitions:** "Rep. Leigh Finke Creates Legislation to Improve Minnesota Human Rights Act," *Legislative News and Views*, March 7, 2023, www.house.mn.gov/members/Profile/News/15575/36641.

77 **Gays Against Groomers:** "Gays Against Groomers," Southern Poverty Law Center, www.splcenter.org/fighting-hate/extremist-files/group/gays-against-groomers.

77 **The group misrepresented Leigh's bill:** Loreben Tuquero, "Minnesota Bill Does Not Create a Protected Class for Child Sex Offenders," *PolitiFact,* May 3, 2023, www.politifact.com/factchecks/2023/may/03/tweets/minnesota-bill-does-not-create-a-protected-class-f.

78 **"Who is served by all this thick":** Roxane Gay, "Jada Pinkett Smith Shouldn't Have to 'Take a Joke.' Neither Should You," *New York Times,* March 29, 2022,

www.nytimes.com/2022/03/29/opinion/culture/will-smith-oscars-roxane-gay.html.

78 **Intersectionality—a popular framework:** Kimberlé W. Crenshaw, "Mapping the Margins: Intersectionality, Identity Politics, and Violence against Women of Color," *Stanford Law Review* 43, no. 6 (1991): 1241–99, www.jstor.org/stable/1229039.

79 **more trauma-informed social media:** Carol F. Scott et al., "Trauma-Informed Social Media: Towards Solutions for Reducing and Healing Online Harm," *CHI '23: Proceedings of the 2023 CHI Conference on Human Factors in Computing Systems*, April 2023, doi.org/10.1145/3544548.3581512.

79 **principles of the trauma-informed approach:** "SAMHSA's Concept of Trauma and Guidance for a Trauma-Informed Approach," SAMHSA's Trauma and Justice Strategic Initiative, July 2014, ncsacw.acf.hhs.gov/userfiles/files/SAMHSA_Trauma.pdf.

80 **disproportionate number of traumatic events:** Jennifer M. Gómez, Robyn L. Gobin, and Melissa L. Barnes, "Discrimination, Violence, and Healing within Marginalized Communities," *Journal of Trauma and Dissociation* 22, no. 2 (2021): 135–40, doi.org/10.1080/15299732.2021.1869059.

80 **Women experience PTSD:** Miranda Olff, "Sex and Gender Differences in Post-Traumatic Stress Disorder: An Update," *European Journal of Psychotraumatology* 8, supp. 4 (2017), doi.org/10.1080/20008198.2017.1351204.

80 **More than half of women:** "Gender-Based Violence and the Effects on Behavioral Health," SAMHSA, www.samhsa.gov/gender-based-violence-behavioral-health.

80 **sexual assault carries a high risk:** Emily R. Dworkin, Anna E. Jaffe, Michele Bedard-Gilligan, and Skye Fitzpatrick, "PTSD in the Year Following Sexual Assault: A Meta-Analysis of Prospective Studies," *Trauma, Violence, and Abuse* 24, no. 2 (2023): 497–514, doi.org/10.1177/15248380211032213.

80 **People who identify as LGBTQ:** M. Marchi et al., "Post-traumatic Stress Disorder Among LGBTQ People: A Systematic Review and Meta-Analysis," *Epidemiology and Psychiatric Sciences* 32 (2023): e44, doi.org/10.1017/S2045796023000586.

80 **Transgender people:** Andrew R. Flores, Ilan Meyer, Lynn L. Langton, and Jody L. Herman, "Gender Identity Disparities in Criminal Victimization: National Crime Victimization Survey 2017–2018," *American Journal of Public Health* 111 (2021): 726–29, doi.org/10.2105/AJPH.2020.306099.

82 **a video of Okamoto dancing:** "@itsaugustco #periodfairy has a whole squad of #periodpixies," Nadya Okamoto, TikTok, www.tiktok.com/@nadyaokamoto/video/7144144294432001326?_t=8n780lKLN0B&_r=1.

82 **"blood coming out of her eyes":** Holly Yan, "Donald Trump's 'Blood' Comment About Megyn Kelly Draws Outrage," CNN, August 8, 2015, www.cnn.com/2015/08/08/politics/donald-trump-cnn-megyn-kelly-comment/index.html.

83 **drinking and rape:** Jaclyn Friedman, "Drinking and Rape: Let's Wise Up About It," *Women's eNews*, February 28, 2007, womensenews.org/2007/02/drinking-and-rape-lets-wise-about-it.

86 **"any strategy for managing a stressful situation":** "Avoidance Coping," *APA Dictionary of Psychology*, American Psychological Association, dictionary.apa .org/avoidance-coping.

Chapter 5: Harm

90 **helicopter crashed:** Scott Cacciola, "Kobe Bryant, Transformational Star of the N.B.A., Dies in Helicopter Crash," *New York Times,* January 26, 2020, www.nytimes.com/2020/01/26/sports/basketball/kobe-bryant-helicopter -crash.html.

90 **the makings of a prodigy:** Marc Stein, "Kobe Bryant Saw His Greatness Mirrored in Gianna," *New York Times,* January 28, 2020, www.nytimes .com/2020/01/28/sports/basketball/kobe-bryant-gianna-shammgod.html.

90 **raping a nineteen-year-old:** Mike Wise and Alex Markels, "Kobe Bryant Charged with Felony Sexual Assault," *New York Times,* July 18, 2003, www .nytimes.com/2003/07/18/sports/basketball/kobe-bryant-charged-with -felony-sexual-assault.html; Marlow Stern, "Kobe Bryant's Disturbing Rape Case: The DNA Evidence, the Accuser's Story, and the Half-Confession," *Daily Beast,* April 11, 2016, www.thedailybeast.com/kobe-bryants-disturbing -rape-case-the-dna-evidence-the-accusers-story-and-the-half-confession.

90 **attacked the character:** Kevin Draper, "Kobe Bryant and the Sexual Assault Case That Was Dropped but Not Forgotten," *New York Times,* January 27, 2020, www.nytimes.com/2020/01/27/sports/basketball/kobe-bryant-rape -case.html.

90 **death threats:** "FBI: Man Threatened to Kill Accuser, Hurlbert," February 27, 2004, www.espn.com/nba/news/story?id=1745738.

90 **dropped the case:** Kirk Johnson, "Prosecutors Drop Kobe Bryant Rape Case," *New York Times,* September 2, 2004, www.nytimes.com/2004/09/02/us /prosecutors-drop-kobe-bryant-rape-case.html.

91 **"as a price of freedom":** "Kobe Case Dismissed at Request of Prosecution," ESPN, September 1, 2004, www.espn.com/nba/news/story?id=1872740.

91 **the statement was negotiated:** "Lawyers Tell How and Why the Accuser of Kobe Bryant Settled for His Apology," *Tampa Bay Times,* November 7, 2004, www.tampabay.com/archive/2004/11/07/lawyers-tell-how-and-why-the -accuser-of-kobe-bryant-settled-for-his-apology.

91 **"Although I truly believe":** "Kobe Bryant's Apology," ESPN, September 1, 2004, www.espn.com/nba/news/story?id=1872928.

91 **out of line:** Paul Farhi, "Washington Post Clears Reporter Who Tweeted Link to Kobe Bryant Rape Allegations," *Washington Post,* January 28, 2020, www .washingtonpost.com/lifestyle/style/washington-post-clears-reporter-who -tweeted-link-to-kobe-bryant-rape-allegations/2020/01/28/74728a32-421a -11ea-b5fc-eefa848cde99_story.html.

91 **provocative clips:** David Bauder, "King Angry at CBS Promo of Interview Questions About Bryant," Associated Press, February 7, 2020, apnews.com /article/gayle-king-ap-top-news-music-sports-general-celebrities-e125bb8d a67b47135fb96d6067fad75f.

91 **false accusations of sexual violence:** Richard Pérez-Peña, "Woman Linked to

1955 Emmett Till Murder Tells Historian Her Claims Were False," *New York Times,* January 27, 2017, www.nytimes.com/2017/01/27/us/emmett-till-lynching-carolyn-bryant-donham.html.

91 **Gayle King:** "Lisa Leslie Reflects on the Legacy of Kobe Bryant," CBS News, February 4, 2020, www.cbsnews.com/news/lisa-leslie-reflects-legacy-kobe-bryant.

92 **was attacked online:** David Bauder, "CBS News Head Calls Threats Against Gayle King Reprehensible," Associated Press, February 8, 2020, apnews.com/article/us-news-ap-top-news-susan-zirinsky-sports-general-gayle-king-1b8a3ad60aa2728fa721fb69e4af4b63.

92 **"It's incoherent":** Amira Rose Davis, "A Legacy of Incoherence," *New Republic*, February 1, 2020, newrepublic.com/article/156398/legacy-incoherence.

92 **"Thanks to the pressures":** Evette Dionne, "We Can Only Process Kobe Bryant's Death by Being Honest About His Life," *Time*, January 28, 2020, time.com/5773151/kobe-bryant-rape-case-complicated-legacy.

92 **I wrote a story:** Alia E. Dastagir, "F-bombs and Death Threats: Kobe Bryant, Mike Pompeo and the Abuse of Women Journalists," *USA Today,* January 27, 2020, www.usatoday.com/story/news/nation/2020/01/27/kobe-bryant-mike-pompeo-mary-louise-kelly-felicia-sonmez/4588915002.

92 **"to ignore that":** Nancy Armour, "Opinion: Don't Shy Away from the Complicated Part of Kobe Bryant's Legacy," *USA Today,* January 27, 2020, www.usatoday.com/story/sports/columnist/nancy-armour/2020/01/27/kobe-bryant-dont-shy-away-nba-legends-legacy/4584605002.

94 **Women who cover sports:** Sarah Guinee, "Dark Side of Sports Journalism as Fans Harass Female Reporters Online," Committee to Protect Journalists, February 28, 2019, cpj.org/2019/02/sports-fans-harass-female-journalist-online.

95 **misgendering trans:** S. E. James et al., *The Report of the 2015 U.S. Transgender Survey* (Washington, DC: National Center for Transgender Equality, 2016), transequality.org/sites/default/files/docs/usts/USTS-Full-Report-Dec17.pdf.

95 **leading to depression:** Nicholas J. Parr and Bethany Grace Howe, "Heterogeneity of Transgender Identity Nonaffirmation Microaggressions and Their Association with Depression Symptoms and Suicidality Among Transgender Persons," *Psychology of Sexual Orientation and Gender Diversity* 6, no. 4 (2019): 461–74, doi.org/10.1037/sgd0000347.

95 **anxiety:** Kai Jacobsen et al., "Misgendering and the Health and Wellbeing of Nonbinary People in Canada," *International Journal of Transgender Health* (2023), doi.org/10.1080/26895269.2023.2278064.

95 **and trauma:** Sebastian M. Barr, Kate E. Snyder, Jill L. Adelson, and Stephanie L. Budge, "Posttraumatic Stress in the Trans Community: The Roles of Anti-Transgender Bias, Non-Affirmation, and Internalized Transphobia," *Psychology of Sexual Orientation and Gender Diversity* 9, no. 4 (2022): 410–21, doi.org/10.1037/sgd0000500.

95 **"If you're exposed to verbal":** Lisa Feldman Barrett, *Seven and a Half Lessons About the Brain* (Boston: Houghton Mifflin Harcourt, 2020), 92.

95 **stress, anxiety, and panic attacks:** "Amnesty Reveals Alarming Impact of Online Abuse Against Women," Amnesty International, November 20, 2017, www.amnesty.org/en/latest/press-release/2017/11/amnesty-reveals-alarming-impact-of-online-abuse-against-women.

95 **literature review:** Francesca Stevens, Jason R.C. Nurse, and Budi Arief, "Cyber Stalking, Cyber Harassment, and Adult Mental Health: A Systematic Review," *Cyberpsychology, Behavior, and Social Networking* 24, no. 6 (2021): 367–76, doi.org/10.1089/cyber.2020.0253.

96 **"economic vandalism":** Emma A. Jane, "Gendered Cyberhate as Workplace Harassment and Economic Vandalism," *Feminist Media Studies* 18 (2018): 1–17, doi.org/10.1080/14680777.2018.1447344.

97 **"toxic speech":** Lynne Tirrell, "Toxic Speech: Toward an Epidemiology of Discursive Harm," *Philosophical Topics* 45, no. 2 (2017): 139–62, www.jstor.org/stable/26529441.

97 **"comes in many varieties":** Lynne Tirrell, "Toxic Speech: Inoculations and Antidotes," *Southern Journal of Philosophy* 56, no. S1 (2018): 116–44, doi.org/10.1111/sjp.12297.

98 **"without awareness of the cumulative effect":** Lynne Tirrell, "Toxic Misogyny and the Limits of Counterspeech," *Fordham Law Review* 6, no. 87 (2019): 2433–52, ir.lawnet.fordham.edu/cgi/viewcontent.cgi?article=5607&context=flr.

98 **susceptibility of the subject:** Tirrell, "Toxic Speech: Toward an Epidemiology."

98 **"We must attend to speech":** Ibid.

99 **we censor ourselves:** Kalyani Chadha, Linda Steiner, Jessica Vitak, and Zahra Ashktorab, "Women's Responses to Online Harassment," *International Journal of Communication* 14, no. 1 (2020): 239–57, ijoc.org/index.php/ijoc/article/view/11683/2906.

99 **ages 15 to 29 censor themselves:** Amanda Lenhart, Michele Ybarra, Kathryn Zickuhr, and Myeshia Price-Feeney, "Online Harassment, Digital Abuse, and Cyberstalking in America," Data and Society Research Institute and Center for Innovative Public Health Research, November 21, 2016, www.datasociety.net/pubs/oh/Online_Harassment_2016.pdf.

99 **"digital Spiral of Silence":** Candi Carter Olson and Victoria LaPoe, "Combating the Digital Spiral of Silence: Academic Activists Versus Social Media Trolls," in *Mediating Misogyny*, ed. J. Vickery and T. Everbach (London: Palgrave Macmillan, 2018), doi.org/10.1007/978-3-319-72917-6_14.

99 **two out of three women journalists:** Michelle Ferrier, "Attacks and Harassment: The Impact on Female Journalists and Their Reporting," TrollBusters and International Women's Media Foundation, 2018, www.iwmf.org/wp-content/uploads/2018/09/Attacks-and-Harassment.pdf.

99 **"Unless you've been there":** Margaret Sullivan, "Online Harassment of Female Journalists Is Real, and It's Increasingly Hard to Endure," *Washington Post*, March 14, 2021, www.washingtonpost.com/lifestyle/media/online-harassment-female-journalists/2021/03/13/ed24b0aa-82aa-11eb-ac37-4383f7709abe_story.html.

100 **Violence online is a social problem:** "PEN America Position Statement: Online Harassment and Free Expression," *Online Harassment Field Manual*,

PEN America, onlineharassmentfieldmanual.pen.org/pen-america-position-statement.

100 **"fucking bitch":** Lisa Lerer, "That Word," *New York Times,* July 23, 2020, www.nytimes.com/2020/07/23/us/politics/aoc-women-ted-yoho.html.

100 **"In front of reporters":** "Alexandria Ocasio-Cortez Speaks Out After Republican's Sexist Attack," *Guardian News,* YouTube, July 23, 2020, www.youtube.com/watch?v=gkQ4DBvjXw8.

103 **publicly tweeting about societal sexism:** Mindi D. Foster, "Tweeting About Sexism: The Well-Being Benefits of a Social Media Collective Action," *British Journal of Social Psychology* 54, no. 4 (2015): 629–47, scholars.wlu.ca/cgi/viewcontent.cgi?article=1050&context=psyc_faculty.

103 **mobilizing support for change:** Ibid.

103 **PSA about women's online abuse:** "#MoreThanMean—Women in Sports 'Face' Harassment," Just Not Sports, YouTube, April 26, 2016, /www.youtube.com/watch?v=9tU-D-m2JY8.

104 **prohibit reporters from speaking:** "Brief of Amicus Curiae Claire Goforth in Support of Appellant/Cross-Appellee in Favor of Reversing the Order of the Superior Court," www.dccourts.gov/sites/default/files/oral-arguments/Brief%20of%20Amici%20Curiae%20Goforth%2022-CV-274.pdf.

104 **favored over censorship:** Lynne Tirrell, "Toxic Misogyny and the Limits of Counterspeech," *Fordham Law Review* 87 (2019): 2433, ir.lawnet.fordham.edu/flr/vol87/iss6/6.

104 **defines counterspeech as:** Counterspeech, The Dangerous Speech Project, dangerousspeech.org/counterspeech.

105 **defined counterspeech broadly:** Bianca Cepollaro, Maxime Lepoutre, and Robert Mark Simpson, "Counterspeech," *Philosophy Compass* 18, no. 1 (2023): e12890, doi.org/10.1111/phc3.12890.

105 **"the form of assertions":** Ibid.

105 **takes on distinct dimensions:** Bianca Cepollaro, "Blocking Toxic Speech Online," in *Conversations Online*, ed. Patrick Connolly, Sanford Goldberg, and Jennifer Saul (New York: Oxford University Press, forthcoming).

106 **not all counterspeakers share the same:** "A Toolkit on Using Counterspeech to Tackle Online Hate Speech," Future of Free Speech, n.d., futurefreespeech.org/a-toolkit-on-using-counterspeech-to-tackle-online-hate-speech.

106 **Some people use counterspeech:** Cathy Buerger, "Why They Do It: Counterspeech Theories of Change," The Dangerous Speech Project, September 26, 2022, dangerousspeech.org/wp-content/uploads/2022/10/Why-They-Do-It-Counterspeech-Theories-of-Change.pdf.

107 **"Where inequality reigns":** Tirrell, "Toxic Misogyny."

107 **senator Marsha Blackburn:** Alia E. Dastagir, "Marsha Blackburn Asked Ketanji Brown Jackson to Define 'Woman.' Science Says There's No Simple Answer," *USA Today,* March 24, 2022, www.usatoday.com/story/life/health-wellness/2022/03/24/marsha-blackburn-asked-ketanji-jackson-define-woman-science/7152439001.

108 **Trump banned all refugees:** Amy Davidson Sorkin, "Trump's Divisive New Travel Ban," *New Yorker*, March 6, 2017, www.newyorker.com/news/amy-davidson/trumps-divisive-new-travel-ban.

108 **Nancy wrote a column:** Nancy Armour, "Opinion: Tom Brady Has Gotten an Undeserved Pass for His Past Support of Donald Trump," *USA Today,* February 2, 2021, www.usatoday.com/story/sports/columnist/nancy-armour/2021/02/02/super-bowl-2021-tom-brady-doesnt-deserve-pass-trump-support/4358829001.

Chapter 6: Trauma

109 **"most harassed technology journalist in America":** Emily Dreyfuss, "What the Harassment of Journalist Taylor Lorenz Can Teach Newsrooms," *Media Manipulation Casebook,* Technology and Social Change Project, March 4, 2022, mediamanipulation.org/research/what-harassment-journalist-taylor-lorenz-can-teach-newsrooms.

109 **some of the richest men on the planet:** Taylor Lorenz, "You Don't Know Taylor Lorenz," *Marie Claire,* October 17, 2023, www.marieclaire.com/culture/taylor-lorenz-essay.

109 **Tucker Carlson has belittled her:** Teo Armus, "Tucker Carlson Keeps Attacking a New York Times Reporter After the Paper Calls His Tactics 'Calculated and Cruel,'" *Washington Post,* March 11, 2021, www.washingtonpost.com/nation/2021/03/11/tucker-carlson-taylor-lorenz-fox.

110 **She has been harassed, abused:** "Gender-Based Online Violence Spikes After Prominent Media Attacks," NYU's Center for Social Media and Politics, January 26, 2022, csmapnyu.org/news-views/news/gender-based-online-violence-spikes-after-prominent-media-attacks.

110 **threatened, mobbed:** Lorenz, "You Don't Know Taylor."

110 **swatted, and stalked:** Dreyfuss, "Lorenz Can Teach Newsrooms."

110 **smear campaigns:** Ibid.

110 **A correction on one of her stories:** Taylor Lorenz, "Who Won the Depp-Heard Trial? Content Creators That Went All-In," *Washington Post,* June 2, 2022, www.washingtonpost.com/technology/2022/06/02/johnny-depp-trial-creators-influencers.

112 **introduced her:** Rebecca Smith, "Tech Journalist Taylor Lorenz on Being 'Extremely Online,' Sponcons Gone Wrong and Her Dream Dinner Party," *dist://ed,* blog.mozilla.org/en/internet-culture/taylor-lorenz-extremely-online-book-interview.

112 **to Tumblr:** Delia Cai, "'I Don't Want to Quit, Because I See What Happens': Taylor Lorenz Still Believes in the Internet," *Vanity Fair,* October 25, 2023, www.vanityfair.com/style/2023/10/taylor-lorenz-extremely-online-interview.

112 **social media at the *Daily Mail*:** "The 60-Second Interview: Taylor Lorenz, Head of Social Media, the Daily Mail/Mail Online," *Politico,* July 28, 2014, www.politico.com/media/story/2014/07/the-60-second-interview-taylor-lorenz-head-of-social-media-the-daily-mail-mail-online-002540.

113 **open fire on two mosques:** Charlotte Graham-McLay, Austin Ramzy and Daniel Victor, "Christchurch Mosque Shootings Were Partly Streamed on Facebook," *New York Times,* March 14, 2019, www.nytimes.com/2019/03/14/world/asia/christchurch-shooting-new-zealand.html.

113 **the gunman was heard:** Graham Macklin, "The Christchurch Attacks:

Livestream Terror in the Viral Video Age," *CTC Sentinel* 12, no. 6 (2019), ctc.westpoint.edu/wp-content/uploads/2019/07/CTC-SENTINEL-062019.pdf.

113 **"Subscribe to PewDiePie":** Kevin Roose, "What Does PewDiePie Really Believe?" *New York Times Magazine*, October 9, 2019, www.nytimes.com/interactive/2019/10/09/magazine/PewDiePie-interview.html.

113 **right-wing provocateur Mike Cernovich:** Charles Homans, "Trumpism's Twitter Wing Celebrates with a 'DeploraBall,'" *New York Times Magazine*, January 20, 2017, www.nytimes.com/2017/01/20/magazine/trumpisms-twitter-wing-celebrates-with-a-deploraball.html.

113 **inciting harassment against women:** "Mike Cernovich," Southern Poverty Law Center, www.splcenter.org/fighting-hate/extremist-files/individual/mike-cernovich.

113 **Jim Hoft:** Maureen O'Connor, "'Smells Like Boys': A Night at the DeploraBall," *New York* magazine, January 20, 2017, www.thecut.com/2017/01/photos-smells-like-boys-a-night-at-the-pro-trump-deploraball.html.

113 **founder of a far-right news site:** Alexis Benveniste, "Twitter Banned Gateway Pundit Founder Jim Hoft," CNN Business, February 7, 2021, www.cnn.com/2021/02/07/media/twitter-ban-gateway-pundit-founder-jim-hoft/index.html.

113 **Gavin McInnes:** Homans, "Trumpism's Twitter Wing Celebrates."

113 **founder of the Proud Boys:** "Proud Boys," Southern Poverty Law Center, www.splcenter.org/fighting-hate/extremist-files/group/proud-boys.

113 **granted White House press credentials:** Michael M. Grynbaum, "White House Grants Press Credentials to a Pro-Trump Blog," *New York Times*, February 13, 2017, www.nytimes.com/2017/02/13/business/the-gateway-pundit-trump.html.

113 **Cernovich deserved to win a Pulitzer:** Inae Oh, "Donald Trump Jr. Wants to Give a Top PizzaGate Conspiracy Theorist a Pulitzer," *Mother Jones*, April 4, 2017, www.motherjones.com/media/2017/04/donald-trump-jr-mike-cernovich.

113 **live-stream the white supremacist:** Brandon Shulleeta, "In Charlottesville and Elsewhere, U.S. Journalists Are Being Assaulted While Covering the News," *Poynter*, August 21, 2017, www.poynter.org/news-release/2017/in-charlottesville-and-elsewhere-u-s-journalists-are-being-assaulted-while-covering-the-news.

113 **gray Dodge Challenger:** *United States v. James Alex Fields, Jr.*, June 27, 2018, www.justice.gov/opa/press-release/file/1075091/download.

113 **James Alex Fields, Jr.:** Ibid.

113 **and injuring nineteen others:** Sheryl Gay Stolberg and Brian M. Rosenthal, "Man Charged After White Nationalist Rally in Charlottesville Ends in Deadly Violence," *New York Times*, August 12, 2017, www.nytimes.com/2017/08/12/us/charlottesville-protest-white-nationalist.html.

113 **punched her in the face:** "Warrant of Arrest—Misdemeanor," U.S. Press Freedom Tracker, August 18, 2017, media.pressfreedomtracker.us/media/documents/Lorenz_assault_courtdocs_redacted_toItEIy.pdf.

114 **Her assailant was later sentenced:** "Journalist Taylor Lorenz Punched While Filming Aftermath of Fatal Attack," U.S. Press Freedom Tracker, November

3, 2017, pressfreedomtracker.us/all-incidents/journalist-taylor-lorenz-punched-while-filming-aftermath-fatal-attack.

114 **tweeted updates:** Robert Farley, "Was Driver Acting in Self-Defense?" *FactCheck*, August 21, 2017, www.factcheck.org/2017/08/driver-acting-self-defense.

114 **gender-based online abuse surged:** Miriam Berger, "Gender-Based Online Abuse Surged During the Pandemic. Laws Haven't Kept Up, Activists Say," *Washington Post,* November 24, 2021, www.washingtonpost.com/world/2021/11/24/online-abuse-surged-during-pandemic-laws-havent-kept-up-activists-say.

114 **prominent male journalists:** "Gender-Based Online Violence," NYU's Center.

114 **tech powerhouses:** Dan Primack, "Tensions Between Tech Industry and Tech Media Boil Over," *Axios*, July 6, 2020, www.axios.com/2020/07/06/tech-industry-media-taylor-lorenz-away-ceo-tensions.

115 **"trade up the chain":** Alice Marwick and Rebecca Lewis, "Media Manipulation and Disinformation Online," Data and Society Research Institute, May 15, 2017, datasociety.net/wp-content/uploads/2017/05/DataAndSociety_MediaManipulationAndDisinformationOnline-1.pdf.

116 **"Destroyed her life, really?":** Armus, "Tucker Carlson Keeps Attacking."

116 **referring to her "giggling":** Shawn McCreesh, "Taylor Lorenz Introduces Her Brand to the Washington Post," *New York Magazine*, March 7, 2022, nymag.com/intelligencer/2022/03/taylor-lorenz-now-at-washington-post-fights-maggie-haberman.html.

116 **"Mean Girl basket [case]":** Kyle Smith, "From Watergate to Whinegate: The Washington Post Is a Hot Mess," *New York Post,* June 6, 2022, nypost.com/2022/06/06/watergate-to-whinegate-the-washington-post-is-a-hot-mess.

116 **"Welcome to the Little":** Tarpley Hitt, "Welcome to the Little Bitch Olympics," *Gawker*, March 8, 2022, www.gawkerarchives.com/media/new-york-magazine-tries-to-bash-taylor-lorenz-fails.

116 **Libs of TikTok:** Taylor Lorenz, "Meet the Woman Behind Libs of TikTok, Secretly Fueling the Right's Outrage Machine," *Washington Post,* April 19, 2022, www.washingtonpost.com/technology/2022/04/19/libs-of-tiktok-right-wing-media.

116 **picked up by conservative news outlets:** Joseph A. Wulfsohn, "Washington Post's Taylor Lorenz Doxxes Libs of TikTok Days After Decrying Online Harassment of Women," Fox News, April 19, 2022, www.foxnews.com/media/washington-post-taylor-lorenz-libs-of-tiktok.

116 **media watchdog Poynter:** Tom Jones, "Opinion: Digging into the Taylor Lorenz Controversy That Should Not Be a Controversy at All," *Poynter*, April 21, 2022, www.poynter.org/newsletters/2022/what-happened-taylor-lorenz-libs-of-tiktok-controversy.

116 **"thin and at times nonexistent":** Whitney Phillips, *This Is Why We Can't Have Nice Things: Mapping the Relationship Between Online Trolling and Mainstream Culture* (Cambridge, MA: MIT Press, 2015), 19.

118 **cognitive avoidance:** Sara J. Sagui-Henson, "Cognitive Avoidance," in *Encyclopedia of Personality and Individual Differences*, ed. V. Zeigler-Hill and

Todd K. Shackelford (Springer, 2017), doi.org/10.1007/978-3-319-28099-8_964-1.

118 **Emotional numbness is characterized:** Or Duek, Rebecca Seidemann, Robert H. Pietrzak, Ilan Harpaz-Rotem, "Distinguishing Emotional Numbing Symptoms of Posttraumatic Stress Disorder from Major Depressive Disorder," *Journal of Affective Disorders* 324 (2023): 294–99, doi.org/10.1016/j.jad.2022.12.105.

118 **associated with trauma:** Brett T. Litz and Matt J. Gray, "Emotional Numbing in Posttraumatic Stress Disorder: Current and Future Research Directions," *Australian and New Zealand Journal of Psychiatry* 36, no. 2 (2002): 198–204, doi.org/10.1046/j.1440-1614.2002.01002.x.

119 **"to protect oneself and to get":** "Demystifying Dissociation: Principles, Best Practices, and Clinical Approaches," www.netce.com/coursecontent.php?courseid=2823#chap.2.

120 **living in violent communities:** Traci M. Kennedy and Rosario Ceballo, "Emotionally Numb: Desensitization to Community Violence Exposure Among Urban Youth," *Developmental Psychology* 52 (2016), doi.org/10.1037/dev0000112.

120 **condemning the attacks:** "IWMF Condemns Online Attacks Against Taylor Lorenz," International Women's Media Foundation, March 2021, www.iwmf.org/2021/03/iwmf-condemns-online-attacks-against-taylor-lorenz.

121 **"In the absence of strong political":** Judith L. Herman, *Trauma and Recovery: The Aftermath of Violence—From Domestic Abuse to Political Terror* (New York: Basic Books, 1992), 17.

Chapter 7: Information Disorder

123 **wrote for the school's paper:** Wilson Conn, "'A Force for Good in Both Our Sport and Our Society': Remembering Grant Wahl '96," *Daily Princetonian*, December 27, 2022, www.dailyprincetonian.com/article/2022/12/remembering-soccer-journalist-grant-wahl-96.

123 **Céline was studying pre-med:** Katherine Hobson, "The Doctor Is On," *Princeton Alumni Weekly,* September 2022, paw.princeton.edu/cbs/doctor/celine/gounder/covid/expert.

123 **Grant was studying politics:** Michael Lewis, "Remembering Grant Wahl," U.S. Soccer, December 20, 2022, www.ussoccer.com/stories/2022/12/remembering-grant-wahl.

123 **honing his skills:** Wilson Conn, "Grant Wahl '96, Revered Sports Journalist and 'Prince' Alum, Dies at 49," *Daily Princetonian*, December 10, 2022, www.dailyprincetonian.com/article/2022/12/grant-wahl-breaking-qatar-world-cup.

124 **There's a picture of them:** "At the wedding of our friends Benji & Carolyn @DrJasik in 2000," Céline Gounder (@celinegounder), Twitter, December 12, 2022, twitter.com/celinegounder/status/1602281128395788288?lang=en.

124 **advocate for women's soccer:** Louisa Thomas, "How Grant Wahl Changed the Place of Soccer in America," *New Yorker*, December 17, 2022, www.newyorker

.com/news/postscript/how-grant-wahl-changed-the-place-of-soccer-in-america.

124 **Grant traveled to Qatar:** Ronald Blum, "US Soccer Journalist Grant Wahl Dies at World Cup," Associated Press, December 10, 2022, apnews.com/article/journalist-grant-wahl-dies-in-qatar-covering-world-cup-82829df66ec0635bdbba3e8edccea96d.

124 **Grant collapsed:** Kevin Draper and Alan Blinder, "Soccer Journalist Dies at World Cup After Collapsing at Argentina Game," *New York Times,* December 9, 2022, www.nytimes.com/2022/12/09/sports/soccer/grant-wahl-dead.html.

124 **Emergency service workers:** Draper and Blinder, "Soccer Journalist Dies."

124 **strangers blamed Grant's death:** Graph Massara, "World Cup Reporter's Fatal Heart Condition Unrelated to COVID Vaccine," Associated Press, December 14, 2022, apnews.com/article/fact-check-grant-wahl-vaccine-false-550141070405.

124 **to link Grant's death to myocarditis:** Céline Gounder, "Grant Wahl Was a Loving Husband. I Will Always Protect His Legacy," *New York Times,* January 8, 2023, www.nytimes.com/2023/01/08/opinion/grant-wahl-celine-gounder-vaccine.html.

124 **a rare side effect:** Massara, "World Cup Reporter's Fatal Heart Condition."

124 **"Now you understand":** Gounder, "Wahl Was a Loving Husband."

124 **released a statement:** "A Note from Grant's wife, Céline Gounder," *Fútbol with Grant Wahl*, December 14, 2022, grantwahl.substack.com/p/a-note-from-grants-wife-celine-gounder.

124 ***The New York Times*:** Apoorva Mandavilli and Andrew Das, "Grant Wahl Died of a Burst Blood Vessel, His Family Says," *New York Times,* December 14, 2022, www.nytimes.com/2022/12/14/health/grant-wahl-death.html.

124 **CBS:** "Dr. Céline Gounder on Late Husband Grant Wahl's Legacy," CBS News, December 14, 2022, www.cbsnews.com/video/dr-celine-gounder-on-late-husband-grant-wahls-legacy.

124 **NPR:** Michel Martin, "Grant Wahl's Wife Remembers the Late Soccer Journalist," *All Things Considered*, NPR, December 17, 2022, www.npr.org/2022/12/17/1143901925/grant-wahls-wife-remembers-the-late-soccer-journalist.

125 **Grant's memorial was on:** Ronald Blum, "Grant Wahl's Life Celebrated at New York City Gathering," Associated Press, December 21, 2022, apnews.com/article/world-cup-sports-soccer-new-york-city-qatar-1ea6d39fd196cad1f37ca63ffb2f5585.

125 **Buffalo Bills safety Damar:** Mitch Stacy, "Bills' Hamlin in Critical Condition After Collapse on Field," Associated Press, January 3, 2023, apnews.com/article/damar-hamlin-collapse-buffalo-bills-cincinnati-bengals-c9f684bdaccd1e3f77bda6c77baca75e.

125 **anti-vaccination community again grew:** Ali Swenson, David Klepper, and Sophia Tulp, "Hamlin's Collapse Spurs New Wave of Vaccine Misinformation," Associated Press, January 4, 2023, apnews.com/article/buffalo-bills-nfl-sports-health-damar-hamlin-b1273c5903a1efd3f67c8480feb6b52f.

125 **"When disinformation profiteers":** Gounder, "Wahl Was a Loving Husband."

126 **agents of disinformation pretending:** Soo Youn, "Black Women Are Being Targeted in Misinformation Campaigns, a Report Shows. Here's What to Know," *Washington Post,* September 30, 2020, www.washingtonpost.com/gender-identity/black-women-are-being-targeted-in-misinformation-campaigns-a-report-shows-heres-what-to-know.

126 **online photos:** Katie Polglase, Pallabi Munsi, Barbara Arvanitidis, Alex Platt, Mark Baron, and Oscar Featherstone, "'My Identity Is Stolen': Photos of European Influencers Used to Push Pro-Trump Propaganda on Fake X Accounts," CNN, August 28, 2024, www.cnn.com/2024/08/28/europe/fake-maga-accounts-x-european-influencers-intl-cmd/index.html.

126 **Deepfake technology:** "2023 State of Deepfakes: Realities, Threats, and Impact," Security Hero, www.securityhero.io/state-of-deepfakes.

127 **Disinformation is content:** Claire Wardle, "Understanding Information Disorder," *First Draft,* September 22, 2020, firstdraftnews.org/long-form-article/understanding-information-disorder.

127 ***Gender disinformation*****:** "Best Practice Forum on Gender and Digital Rights: Exploring the Concept of Gendered Disinformation," Internet Governance Forum, intgovforum.org/en/filedepot_download/248/21181.

127 **"false or misleading":** "Gendered Disinformation: Tactics, Themes, and Trends by Foreign Malign Actors," U.S. Department of State, March 27, 2023, www.state.gov/gendered-disinformation-tactics-themes-and-trends-by-foreign-malign-actors.

127 **a piece about victim blaming:** Jaclyn Friedman, "Drinking and Rape: Let's Wise Up About It," Women's eNews, February 28, 2007, womensenews.org/2007/02/drinking-and-rape-lets-wise-about-it.

127 **online manifesto:** Zoë Quinn, *Crash Override: How Gamergate (Nearly) Destroyed My Life, and How We Can Win the Fight Against Online Hate* (New York: PublicAffairs, 2017), 1–7.

128 **went on Instagram Live:** Alexandria Ocasio-Cortez (aoc), February 2, 2021, www.instagram.com/p/CKxlyx4g-Yb.

128 **clearly stated:** Ali Swenson, "Ocasio-Cortez Didn't Lie About Location During Capitol Riot," Associated Press, February 4, 2021, apnews.com/article/fact-checking-9951968706.

128 **didn't act on 95 percent:** "Failure to Act: How Tech Giants Continue to Defy Calls to Rein in Vaccine Misinformation," Center for Countering Digital Hate, December 1, 2020, counterhate.com/wp-content/uploads/2022/05/201201-Failure-to-Act.pdf.

129 ***welfare queen*****:** Alice Marwick, Rachel Kuo, Shanice Jones Cameron, and Moira Weigel, "Critical Disinformation Studies: A Syllabus," Center for Information, Technology, and Public Life, citap.unc.edu/wp-content/uploads/2023/02/Marwick_Kuo_Cameron_Weigel_2021_CriticalDisinformationStudiesSyllabus.pdf.

129 **homophobic disinformation:** Evelyn Schlatter and Robert Steinback, "10 Anti-Gay Myths Debunked," Intelligence Report, Southern Poverty Law Center, February 27, 2011, www.splcenter.org/fighting-hate/intelligence-report/2011/10-anti-gay-myths-debunked.

129 **Kamala Harris:** Karen Tumulty, Kate Woodsome, and Sergio Peçanha, "How Sexist, Racist Attacks on Kamala Harris Have Spread Online—A Case Study," *Washington Post,* October 7, 2020, www.washingtonpost.com/opinions/2020/10/07/kamala-harris-sexist-racist-attacks-spread-online/?arc404=true.

129 **worked as a prostitute:** Michael Gold, "Trump Laughs as Supporter Yells That Harris 'Worked on the Corner,'" *New York Times,* November 3, 2024, www.nytimes.com/2024/11/03/us/politics/trump-harris-crude-remark.html.

130 **targeted women:** Youn, "Black Women Are Being Targeted."

130 **"Black identity and culture":** Shireen Mitchell, "How the Facebook Ads That Targeted Voters Centered on Black American Culture: Voter Suppression Was the End Game," Stop Online Violence Against Women, October 11, 2018, stoponlinevaw.com/wp-content/uploads/2018/10/Black-ID-Target-by-Russia-Report-SOVAW.pdf.

130 **spread misleading vaccine information:** Sheera Frenkel, "Black and Hispanic Communities Grapple with Vaccine Misinformation," *New York Times,* March 10, 2021, www.nytimes.com/2021/03/10/technology/vaccine-misinformation.html.

130 **drive a wedge between them:** Patricia Mazzei and Jennifer Medina, "False Political News in Spanish Pits Latino Voters Against Black Lives Matter," *New York Times,* October 21, 2020, www.nytimes.com/2020/10/21/us/politics/spanish-election-2020-disinformation.html.

130 **long experienced abuse and exploitation:** Alia E. Dastagir, "Black Americans' Health Is in Crisis. What Will It Take for Them to Be Well?" *USA Today,* February 8, 2022, www.usatoday.com/in-depth/life/health-wellness/2022/02/08/racism-and-discrimination-remain-dangerous-black-health-wellness/9091616002.

130 **Black bodies pulled from graves:** Allison C. Meier, "Grave Robbing, Black Cemeteries, and the American Medical School," *JSTOR Daily*, August 24, 2018, daily.jstor.org/grave-robbing-black-cemeteries-and-the-american-medical-school.

130 **Black women sterilized without their:** Linda Villarosa, "The Long Shadow of Eugenics in America," *New York Times Magazine*, June 8, 2022, www.nytimes.com/2022/06/08/magazine/eugenics-movement-america.html.

130 **notorious Tuskegee syphilis experiment:** Jean Heller, "Black Men Untreated in Syphilis Study," Associated Press, July 25, 1972, apnews.com/article/business-science-health-race-and-ethnicity-syphilis-e9dd07eaa4e74052878a68132cd3803a.

131 **women of color candidates:** Dhanaraj Thakur and DeVan Hankerson Madrigal, "An Unrepresentative Democracy: How Disinformation and Online Abuse Hinder Women of Color Political Candidates in the United States," Center for Democracy and Technology, October 27, 2022, cdt.org/insights/an-unrepresentative-democracy-how-disinformation-and-online-abuse-hinder-women-of-color-political-candidates-in-the-united-states.

131 **"their families also":** Lucina Di Meco, "Monetizing Misogyny: Gendered Disinformation and the Undermining of Women's Rights and Democracy

Globally," #ShePersisted, February 2023, she-persisted.org/wp-content/uploads/2023/02/ShePersisted_MonetizingMisogyny.pdf.

131 **93 percent remained:** "Abusing Women in Politics: How Instagram Is Failing Women and Public Officials," Center for Countering Digital Hate, August 2024, counterhate.com/wp-content/uploads/2024/08/Abusing-Women-In-Politics_CCDH_Int.pdf.

131 **an Israeli-based group:** Michael McGowan, Christopher Knaus, and Nick Evershed, "Monetising Hate: Covert Enterprise Co-Opts Far-Right Facebook Pages to Churn Out Anti-Islamic Posts," *Guardian,* December 5, 2019, www.theguardian.com/technology/2019/dec/05/monetising-hate-covert-enterprise-co-opts-far-right-facebook-pages-to-churn-out-anti-islamic-posts.

131 **targeted Muslim lawmakers:** David Smith, Michael McGowan, Christopher Knaus, and Nick Evershed, "Revealed: Ilhan Omar and Rashida Tlaib Targeted in Far-Right Fake News Operation," *Guardian,* December 5, 2019, www.theguardian.com/technology/2019/dec/05/ilhan-omar-rashida-tlaib-targeted-far-right-fake-news-operation-facebook.

131 **first Muslim women to serve in Congress:** Michelle Boorstein, Marisa Iati, and Julie Zauzmer Weil, "The Nation's First Two Muslim Congresswomen Are Sworn In, Surrounded by the Women They Inspired," *Washington Post,* January 3, 2019, www.washingtonpost.com/religion/2019/01/03/americas-first-two-muslim-congresswomen-are-sworn-surrounded-by-women-they-inspired.

132 **"placed ads from major brands":** Craig Silverman, Ruth Talbot, Jeff Kao, and Anna Klühspies, "How Google's Ad Business Funds Disinformation Around the World," *ProPublica*, October 29, 2022, www.propublica.org/article/google-alphabet-ads-fund-disinformation-covid-elections.

132 **"thoughts and feelings":** "Compartmentalization," *APA Dictionary of Psychology,* American Psychological Association, dictionary.apa.org/compartmentalization.

133 **a racist remark:** Negar Fani et al., "Association of Racial Discrimination with Neural Response to Threat in Black Women in the US Exposed to Trauma," *JAMA Psychiatry* 78, no. 9 (2021): 1005–12, jamanetwork.com/journals/jamapsychiatry/fullarticle/2782454.

134 **"separates various aspects of the self":** Vera Békés, Yocheved Ayden Ferstenberg, and J. Christopher Perry, "Compartmentalization," in *Encyclopedia of Personality and Individual Differences*, ed. Virgil Zeigler-Hill and Todd K. Shackelford (Springer, 2020).

134 **splitting:** Ryan Bailey and Jose Pico, "Defense Mechanisms," *StatPearls*, May 22, 2023, www.ncbi.nlm.nih.gov/books/NBK559106/.

134 **repression:** Ibid.

134 **dissociation defenses:** Ibid.

Chapter 8: Online, Offline, What Line?

138 **"Being able to leave":** Sara Ahmed, *Living a Feminist Life* (Durham, NC: Duke University Press: 2017), 244.

140 **"Fucking bitch":** Sarah N. Lynch, "U.S. Congresswoman Ocasio-Cortez Says Republican Colleague Called Her Profane Slur," Reuters, July 23, 2020, www.reuters.com/article/idUSKCN24O2F3.

141 **"We wouldn't say it":** "#MoreThanMean—Women in Sports 'Face' Harassment," *Just Not Sports*, April 26, 2016, www.youtube.com/watch?v=9tU-D-m2JY8.

142 **"shove a sock":** Natassia Chrysanthos and Rob Harris, "Alan Jones Tells Scott Morrison to 'Shove a Sock Down Throat' of Jacinda Ardern," *Sydney Morning Herald*, August 15, 2019, www.smh.com.au/national/alan-jones-tells-scott-morrison-to-shove-a-sock-down-throat-of-jacinda-ardern-20190815-p52hja.html.

142 **words Donald Trump used:** Janell Ross, "So Which Women Has Donald Trump Called 'Dogs' and 'Fat Pigs'?" *Washington Post,* August 8, 2015, www.washingtonpost.com/news/the-fix/wp/2015/08/08/so-which-women-has-donald-trump-called-dogs-and-fat-pigs.

142 **to describe us:** Stef W. Kight, "A List of Trump's Attacks on Prominent Women," *Axios*, October 16, 2017, www.axios.com/2017/12/15/a-list-of-trumps-attacks-on-prominent-women-1513303964.

142 **"dog":** Greg Price, "Donald Trump Has a Long History of Calling Women 'Dogs' Before Omarosa," *Newsweek*, August 14, 2018, www.newsweek.com/donald-trump-women-dogs-omarosa-1071808.

142 **"happened to turn Black":** Brett Samuels, "Trump Mocks Harris's Heritage: 'She Happened to Turn Black,'" *The Hill,* July 31, 2024, thehill.com/homenews/campaign/4803783-trump-mocks-harris-heritage.

142 **called her "nasty":** Maggie Haberman and Jonathan Swan, "Inside the Worst Three Weeks of Donald Trump's 2024 Campaign," *New York Times,* August 10, 2024, www.nytimes.com/2024/08/10/us/politics/trump-campaign-election.html.

143 **5.8 percent of welding:** "Labor Force Statistics from the Current Population Survey," U.S. Bureau of Labor Statistics, www.bls.gov/cps/cpsaat11.htm.

143 **damaging radiation:** "Welding—Radiation and the Effects on Eyes and Skin," Canadian Centre for Occupational Health and Safety, www.ccohs.ca/oshanswers/safety_haz/welding/eyes.html#.

143 **2.2 percent of plumbers:** "Labor Force Statistics from the Current Population Survey," U.S. Bureau of Labor Statistics, www.bls.gov/cps/cpsaat11.htm.

144 **handful of women:** "Women's History Month: Brooke Nichols, an Apprentice Plumber in Plumbers Local 200," News 12 Long Island, March 29, 2024, longisland.news12.com/womens-history-month-brooke-nichols-an-apprentice-plumber-in-plumbers-local-200.

145 **"digital technologies do not merely facilitate":** Debbie Ging and Eugenia Siapera, "Special Issue on Online Misogyny," *Feminist Media Studies* 18, no. 4 (2018): 515–24, doi.org/10.1080/14680777.2018.1447345.

146 **body budgets in good shape:** Lisa Feldman Barrett, *How Emotions Are Made: The Secret Life of the Brain* (Boston: Houghton Mifflin Harcourt, 2017), 177.

146 **PEN America's field manual:** "Self-Care," *Online Harassment Field Manual,* PEN America, onlineharassmentfieldmanual.pen.org/self-care.

147 **For some trauma survivors:** Loretta Pyles, *Healing Justice: Holistic Self-Care for Change Makers* (New York: Oxford University Press, 2018), 104.

147 **after experiencing burnout:** Ibid.

148 **Writing therapy was pioneered:** James W. Pennebaker, "Expressive Writing in Psychological Science," *Perspectives on Psychological Science* 13, no. 2 (2018): 226–29, doi.org/10.1177/1745691617707315.

148 **Pennebaker wrote an article:** James W. Pennebaker, "Writing About Emotional Experiences as a Therapeutic Process," *Psychological Science* 8, no. 3 (1997): 162–66, doi.org/10.1111/j.1467-9280.1997.tb00403.x.

148 **Pennebaker would later write:** James W. Pennebaker, "Expressive Writing in Psychological Science," *Perspectives on Psychological Science* 13, no. 2 (2018): 226–29, doi.org/10.1177/1745691617707315.

149 **"sitting on the Staten Island Ferry":** Audre Lorde, "A Burst of Light," in *A Burst of Light and Other Essays* (Ixia Press, 2017), 124.

150 **"because several generations ago":** Clint Smith, *How the Word Is Passed: A Reckoning with the History of Slavery Across America* (Boston: Little, Brown, 2021), 234.

Chapter 9: Women, Too

152 **Hillary Clinton:** "Hillary Rodham Clinton," First Families, White House, www.whitehouse.gov/about-the-white-house/first-families/hillary-rodham-clinton.

152 **Madeleine Albright:** "Biographies of the Secretaries of State: Madeleine Korbel Albright (1937–2022)," Office of the Historian, U.S. Department of State, history.state.gov/departmenthistory/people/Albright-madeleine-korbel.

152 **Nora Ephron:** "Nora Ephron '62 Addressed the Graduates in 1996," Commencement Archives, 1996, Wellesley College, www.wellesley.edu/events/commencement/archives/1996commencement.

152 **"Non Ministrari sed Ministrare":** "About: Mission and Values," Wellesley College, www.wellesley.edu/about/missionandvalues.

152 **record number of women:** Guy Gugliotta, "'Year of the Woman' Becomes Reality as Record Number Win Seats," *Washington Post,* November 4, 1992, www.washingtonpost.com/archive/politics/1992/11/04/year-of-the-woman-becomes-reality-as-record-number-win-seats/77598a23-dea3-491a-a8b3-9ccfaf7b61ef.

152 **first woman secretary of state:** "Biographies of the Secretaries of State: Madeleine Korbel Albright."

152 **Clinton became a U.S. senator:** "Hillary Clinton," White House.

152 **At Wellesley:** "New Alumnae Association Board Members," *Wellesley Magazine,* Summer 2016, magazine.wellesley.edu/summer-2016/new-alumnae-association-board-members.

153 **"Women are first-class at Wellesley":** Alexandra Schwartz, "Waiting for the Female Future at Wellesley," *New Yorker,* November 10, 2016, www.newyorker.com/culture/culture-desk/waiting-for-the-female-future-at-wellesley.

153 **serve as a special adviser:** Office of the Attorney General Letitia James,

"Report of Investigation into Allegations of Sexual Harassment by Governor Andrew M. Cuomo," State of New York (2021), ag.ny.gov/sites/default/files/2021.08.03_nyag_-_investigative_report.pdf.

153 **strip poker:** Lindsey Boylan, "My Story of Working with Governor Cuomo," *Medium*, February 24, 2021, lindseyboylan4ny.medium.com/my-story-of-working-with-governor-cuomo-e664d4814b4e.

153 **publicly accuse Cuomo:** Eric Lach, "Who Ordered a Smear Campaign Against Andrew Cuomo's First Accuser?" *New Yorker*, March 9, 2021, www.newyorker.com/news/our-local-correspondents/who-ordered-a-smear-campaign-against-andrew-cuomos-first-accuser.

154 **two years after she resigned:** Office of James, "Report of Investigation Into Allegations of Sexual Harassment."

154 **running for Manhattan borough president:** Lach, "Who Ordered a Smear Campaign?"

154 **bid for Congress:** Bruce Handy, "How to Run a Grassroots Campaign in a Pandemic," *New Yorker,* June 22, 2020, www.newyorker.com/magazine/2020/06/29/how-to-run-a-grassroots-campaign-in-a-pandemic.

154 **already tweeted:** Lach, "Who Ordered a Smear Campaign?"

154 **popular national figure:** Nick Paumgarten, "Andrew Cuomo, the King of New York," *New Yorker*, October 12, 2020, www.newyorker.com/magazine/2020/10/19/andrew-cuomo-the-king-of-new-york.

154 **Lindsey was in the car:** Ronan Farrow, "Cuomo's First Accuser Raises New Claims of Harassment and Retaliation," *New Yorker,* March 18, 2021, www.newyorker.com/news/news-desk/cuomos-first-accuser-raises-new-claims-of-harassment-and-retaliation.

154 **in a *Medium* post:** Lindsey Boylan, "My Story of Working with Governor Cuomo," *Medium,* February 24, 2021, lindseyboylan4ny.medium.com/my-story-of-working-with-governor-cuomo-e664d4814b4e.

155 **women are almost as likely:** Jamie Bartlett et al., "Misogyny on Twitter," *Demos,* May 2014, demos.co.uk/wp-content/uploads/files/MISOGYNY_ON_TWITTER.pdf.

155 **Group Analytic Society International was formed:** "History," Group Analytic Society, groupanalyticsociety.co.uk/about-gasi/our-history.

156 **interested in the relationship:** "Group Analysis," Group Analytic Society, groupanalyticsociety.co.uk/about-gasi/group-analysis.

156 **to give the annual Foulkes Lecture:** Sue Einhorn, "44th Foulkes Lecture," May 21, 2021," www.youtube.com/watch?v=qCY3OeIk6MY.

156 **relationships between women:** Sue Einhorn, "From a Woman's Point of View: How Internalized Misogyny Affects Relationships Between Women," *Group Analysis* 54, no. 4 (2021): 481–98, doi.org/10.1177/05333164211038310.

156 **other women did, too:** J. David Goodman and Luis Ferré-Sadurní, "'I Believe These 11 Women,' Letitia James Says as She Reveals the Report's Findings," *New York Times,* August 3, 2021, www.nytimes.com/2021/08/03/nyregion/letitia-james-ag.html.

156 **"the Governor engaged in":** Office of James, "Report of Investigation Into Allegations of Sexual Harassment."

157 **Cuomo announced his resignation:** Luis Ferré-Sadurní and J. David Goodman, "Cuomo Resigns Amid Scandals, Ending Decade-Long Run in Disgrace," *New York Times,* August 10, 2021, www.nytimes.com/2021/08/10/nyregion/andrew-cuomo-resigns.html.

157 **three out of four appeared:** Nicholas Fandos and Dana Rubinstein, "Why These Women Are Determined to Clear Cuomo's Name," *New York Times,* February 2, 2022, www.nytimes.com/2022/02/02/nyregion/cuomo-supporters.html.

157 **"Online, they have banded":** Fandos and Rubinstein, "Why These Women."

157 **Madeline Cuomo:** Nicholas Fandos, "The Secret Hand Behind the Women Who Stood by Cuomo? His Sister," *New York Times,* August 7, 2023, www.nytimes.com/2023/08/07/nyregion/cuomo-women-sister-madeline.html.

157 **worked to discredit Lindsey:** Office of James, "Report of Investigation Into Allegations of Sexual Harassment."

157 **works at Facebook's parent:** Dani Lever, LinkedIn profile.

157 **suggested Lindsey Boylan was a liar:** Jesse McKinley and Luis Ferré-Sadurní, "Ex-Aide Details Sexual Harassment Claims Against Gov. Cuomo," *New York Times,* February 24, 2021, www.nytimes.com/2021/02/24/nyregion/cuomo-lindsey-boylan-harassment.html.

158 **"an ideology":** Koa Beck, *White Feminism: From the Suffragettes to the Influencers and Who They Leave Behind* (New York: Atria Books, 2021), xvii.

158 **Former Meta COO Sheryl Sandberg:** Mike Isaac, Sheera Frenkel, and Cecilia Kang, "Sheryl Sandberg Is Stepping Down from Meta," *New York Times,* June 1, 2022, www.nytimes.com/2022/06/01/technology/sheryl-sandberg-facebook.html.

158 ***Lean In* became a bestseller:** "Sandberg's 'Lean In' Tops Bestseller Chart," *Publishers Weekly*, March 21, 2013, www.publishersweekly.com/pw/by-topic/industry-news/publisher-news/article/56475-sandberg-s-lean-in-tops-bestseller-chart.html.

158 **the structural issues:** Vauhini Vara, "Sheryl Sandberg's Divisive Pitch to #leanintogether," *New Yorker*, March 8, 2015, www.newyorker.com/business/currency/sheryl-sandbergs-divisive-pitch-to-leanintogether.

158 **product she helped run:** Nellie Bowles, "Lean In's Sheryl Sandberg Problem," *New York Times,* December 7, 2018, www.nytimes.com/2018/12/07/technology/lean-in-sheryl-sandberg-problem.html.

159 **Clinton, who has also been:** Beck, *White Feminism,* 226.

159 **"difficult to read":** Farrow, "Cuomo's First Accuser."

159 **backed Cuomo for a third term:** Jimmy Vielkind, "Hillary Clinton Endorses Cuomo, Says He's 'Getting Things Done,'" *Politico,* May 23, 2018, www.politico.com/states/new-york/albany/story/2018/05/23/hillary-clinton-endorses-cuomo-says-hes-getting-things-done-434245.

159 **leaked to members of the media:** Office of James, "Report of Investigation Into Allegations of Sexual Harassment."

160 **confrontational encounters:** "Ronan Farrow on Interview with 1st Cuomo Accuser," *Good Morning America,* ABC, March 19, 2021, www.youtube.com/watch?v=pdwq4WMYiX0.

160 **"the layers and layers":** Beck, *White Feminism,* 223.

161 **more than seven in ten survivors:** Jasmine Tucker and Jennifer Mondino, *Coming Forward: Key Trends and Data from the TIME'S UP Legal Defense Fund*, National Women's Law Center and the TIME'S UP Legal Defense Fund, October 13, 2020, nwlc.org/wp-content/uploads/2020/10/NWLC-Intake-Report_FINAL_2020-10-13.pdf.

161 **a panel on Muslims:** Jason McGahan, "On Civil Liberty, a Free Exchange," *Washington Post,* December 19, 2001, www.washingtonpost.com/archive/local/2001/12/20/on-civil-liberty-a-free-exchange/cc46a62d-0d9d-41e2-8f42-6806df04cdbd.

161 **helped gut homes:** "A Year of Helping Hands: Wellesley Sends Volunteers, Interns, Money and Support for Hurricane Disaster Relief" (press release), Office of Public Affairs, Wellesley College, June 9, 2006, web.wellesley.edu/PublicAffairs/Releases/2006/060906.html.

162 **"logotherapy":** Viktor E. Frankl, *Man's Search for Meaning* (Boston: Beacon Press), 98.

162 **a theory Frankl developed:** Ibid., 157.

162 **"There is nothing in the world":** Ibid., 103.

163 **purpose is a component:** Patrick E. McKnight and Todd B. Kashdan, "Purpose in Life as a System That Creates and Sustains Health and Well-Being: An Integrative, Testable Theory," *Review of General Psychology* 13, no. 3 (2009): 242–51, doi.org/10.1037/a0017152.

163 **may add years to a person's life:** Patrick L. Hill and Nicholas A. Turiano, "Purpose in Life as a Predictor of Mortality Across Adulthood," *Psychological Science* 25, no. 7 (2014): 1482–86, doi.org/10.1177/0956797614531799.

163 **can help a person recover:** Stacey M. Schaefer et al., "Purpose in Life Predicts Better Emotional Recovery from Negative Stimuli," *PLoS One* 8, no. 11 (2013): e80329, doi.org/10.1371/journal.pone.0080329.

163 **Two men in the camp:** Frankl, *Man's Search,* 79.

Chapter 10: A Place All May Enter

166 **remains of two hundred children:** Ian Austen, "'Horrible History': Mass Grave of Indigenous Children Reported in Canada," *New York Times,* May 28, 2021, www.nytimes.com/2021/05/28/world/canada/kamloops-mass-grave-residential-schools.html.

166 **Canada's residential schools:** Ibid.

166 **operated by churches:** Ibid.

166 **beaten and sexually abused:** Halle Nelson, "Remembering the Children of Native American Residential Schools," National Sexual Violence Resource Center, November 22, 2022, www.nsvrc.org/blogs/remembering-children-native-american-residential-schools.

166 **tongues pricked with needles:** Kylie Rice, "Residential Schools and Their Lasting Impact," Indigenous Foundation, www.theindigenousfoundation.org/articles/residential-schools-their-lasting-impacts.

166 **Many never returned home:** Ian Austen, "The Indigenous Archaeologist

Tracking Down the Missing Residential Children," *New York Times,* July 30, 2021, www.nytimes.com/2021/07/30/world/canada/indigenous-archaeologist-graves-school-children.html.

167 **missing and murdered Indigenous women:** Taima Moeke-Pickering, Julia Rowat, Sheila Cote-Meek, and Ann Pegoraro, "Indigenous Social Activism Using Twitter: Amplifying Voices Using #MMIWG," in *Indigenous Peoples Rise Up: The Global Ascendency of Social Media Activism*, ed. Bronwyn Carlson and Jeff Berglund (New Brunswick, NJ: Rutgers University Press, 2021).

167 **#idlenomore:** Marisa Elena Duarte, *Network Sovereignty: Building the Internet Across Indian Country* (Seattle: University of Washington Press, 2017), 3–5.

167 **"a divergence between Indigenous worldviews":** Steve Elers, Phoebe Elers, and Mohan Dutta, "Responding to White Supremacy: An Analysis of Twitter Messages by Māori After the Christchurch Terrorist Attack," in *Indigenous Peoples Rise Up: The Global Ascendency of Social Media Activism*, ed. Bronwyn Carlson and Jeff Berglund (News Brunswick, NJ: Rutgers University Press, 2021), 65–79.

167 **"serve to only amplify capitalism":** Leanne Betasamosake Simpson, *As We Have Always Done: Indigenous Freedom through Radical Resistance* (University of Minnesota Press, 2020), 221.

168 **"A Declaration of the Independence":** John Perry Barlow, "A Declaration of the Independence of Cyberspace," Electronic Frontier Foundation, February 8, 1996, www.eff.org/cyberspace-independence.

168 **"Manifest Destiny version 2.0":** Whitney Phillips, *This Is Why We Can't Have Nice Things: Mapping the Relationship between Online Trolling and Mainstream Culture* (Cambridge, MA: MIT Press, 2015), 172.

169 **"echo Barlow's utopian vision":** Ibid.

169 **"spurred by expansionist and colonialist ideologies":** Ibid., 19.

169 **"go further, to go faster":** Ibid., 171.

169 **"underlying the histories we tell":** Adrienne L. Massanari, *Gaming Democracy: How Silicon Valley Leveled Up the Far Right* (Cambridge, MA: MIT Press, 2024), 67.

169 **"Western frontier myths":** Ibid., 71–72.

170 **women were at the start:** Claire L. Evans, *Broad Band: The Untold Story of the Women Who Made the Internet* (New York: Portfolio/Penguin, 2018).

170 **Ada Lovelace:** Ibid., 21.

170 **working on computational projects:** Ibid., 53.

170 **women of the ENIAC 6:** Ibid., 38.

170 **Social Services Referral Directory:** Ibid., 106.

170 **"They all care deeply":** Ibid., 3.

170 **persistent stereotypes:** Emily Chang, *Brotopia: Breaking Up the Boys' Club of Silicon Valley* (New York: Portfolio/Penguin, 2018), 16–41.

170 **had never heard of the Internet:** "Internet & Tech," Pew Charitable Trusts, n.d., www.pewtrusts.org/en/topics/internet-and-tech.

171 **men made up a disproportionate share:** "Online Use," Pew Research Center, December 16, 1996, www.pewresearch.org/politics/1996/12/16/online-use.

171 **Internet use is near ubiquitous:** "Internet, Broadband Fact Sheet," Pew

Research Center, January 31, 2024, www.pewresearch.org/internet/fact-sheet/internet-broadband.

171 **Hot or Not–inspired:** Chang, *Brotopia*, 239.

171 **"prank website":** Alex Horton, "Channeling 'The Social Network,' Lawmaker Grills Zuckerberg on His Notorious Beginnings," *Washington Post,* April 11, 2018, www.washingtonpost.com/news/the-switch/wp/2018/04/11/channeling-the-social-network-lawmaker-grills-zuckerberg-on-his-notorious-beginnings.

171 **Janet Jackson's humiliating Super Bowl:** Jim Hopkins, "Surprise! There's a Third YouTube Co-founder," *USA Today*, October 11, 2006, usatoday30.usatoday.com/tech/news/2006-10-11-youtube-karim_x.htm.

171 **hotbeds for trolls:** Chang, *Brotopia*, 227.

171 **rise to the top:** Andrew Marantz, "The Dark Side of Techno-Utopianism," *New Yorker*, September 23, 2019, www.newyorker.com/magazine/2019/09/30/the-dark-side-of-techno-utopianism.

171 **CEO Jack Dorsey:** "Twitter CEO Says His and Other Tech Firms Have Not Combated Abuse Enough," Reuters, February 12, 2019, www.reuters.com/article/us-twitter-dorsey/twitter-ceo-says-his-and-other-tech-firms-have-not-combated-abuse-enough-idUSKCN1Q202T.

171 **his own brand of so-called free speech:** Kate Conger, Ryan Mac, and Mike Isaac, "Elon Musk Fires Twitter Employees Who Criticized Him," *New York Times,* November 15, 2022, www.nytimes.com/2022/11/15/technology/elon-musk-twitter-fired-criticism.html.

171 **changing course on content moderation:** "Former Twitter Insider Describes Elon Musk's Mixed Signals on Free Speech," *Frontline*, PBS, October 10, 2023, www.pbs.org/wgbh/frontline/article/twitter-elon-musk-free-speech-x-documentary-excerpt.

171 **gutted the company's staff:** Kate Conger, Ryan Mac, and Mike Isaac, "In Latest Round of Job Cuts, Twitter Is Said to Lay Off at Least 200 Employees," *New York Times,* February 26, 2023, www.nytimes.com/2023/02/26/technology/twitter-layoffs.html.

171 **30 percent of its Trust and Safety:** Mark Scott, "After Elon Musk's Takeover, X Slashed Trust and Safety Team by 30 Percent," *Politico Pro*, January 10, 2024, subscriber.politicopro.com/article/2024/01/after-elon-musks-takeover-x-slashed-trust-and-safety-team-by-30-percent-00134755.

171 **reinstated many big accounts:** Taylor Lorenz, "'Opening the Gates of Hell': Musk Says He Will Revive Banned Accounts," *Washington Post,* November 24, 2022, www.washingtonpost.com/technology/2022/11/24/twitter-musk-reverses-suspensions.

171 **"high-profile users":** Jeff Horwitz, "Facebook Says Its Rules Apply to All. Company Documents Reveal a Secret Elite That's Exempt," *Wall Street Journal*, September 13, 2021, www.wsj.com/articles/the-facebook-files-11631713039.

171 **"abuse the privilege":** Ibid.

172 **Facebook leadership nixed the plan:** Elizabeth Dwoskin, Nitasha Tiku, and Craig Timberg, "Facebook's Race-Blind Practices Around Hate Speech Came at the Expense of Black Users, New Documents Show," *Washington Post,*

November 21, 2021, www.washingtonpost.com/technology/2021/11/21/facebook-algorithm-biased-race.

172 **"hold all types of values":** Safyia Noble, *Algorithms of Oppression: How Search Engines Reinforce Racism* (New York: New York University Press, 2018), 1.

172 **Meta, formerly Facebook, was systemically silencing:** "Meta's Broken Promises: Systemic Censorship of Palestine Content on Instagram and Facebook," Human Rights Watch, December 21, 2023, www.hrw.org/report/2023/12/21/metas-broken-promises/systemic-censorship-palestine-content-instagram-and.

172 **Israel has been accused:** Masha Green, "The Limits of Accusing Israel of Genocide," *New Yorker*, February 7, 2024, www.newyorker.com/news/our-columnists/the-limits-of-accusing-israel-of-genocide-under-international-law.

172 **the ICJ ruled:** "Legal Consequences Arising from the Policies and Practices of Israel in the Occupied Palestinian Territory, Including East Jerusalem," July 19, 2024, www.icj-cij.org/sites/default/files/case-related/186/186-20240719-adv-01-00-en.pdf.

172 **"corporations controlled by":** Leanne Betasamosake Simpson, *As We Have Always Done: Indigenous Freedom through Radical Resistance* (University of Minnesota Press, 2020), 222.

173 **"violence that we have to survive":** Sara Ahmed, *Living a Feminist Life* (Durham, NC: Duke University Press, 2017), 80.

173 **highest rates of violence:** "Indigenous Peoples," Amnesty International, n.d., www.amnesty.org/en/what-we-do/indigenous-peoples.

173 **epidemic of violence:** "Missing and Murdered Indigenous People Crisis," U.S. Department of the Interior Indian Affairs, n.d., www.bia.gov/service/mmu/missing-and-murdered-indigenous-people-crisis.

173 **more than one thousand women and girls:** "Missing and Murdered Indigenous Women and Girls: The Facts," Amnesty International, January 29, 2021, www.amnesty.ca/blog/missing-and-murdered-indigenous-women-facts.

173 **The Canadian government reports:** "Reclaiming Power and Place: The Final Report of the National Inquiry into Missing and Murdered Indigenous Women and Girls," vol. 1a, National Inquiry into Missing and Murdered Indigenous Women and Girls, 2019, www.mmiwg-ffada.ca/wp-content/uploads/2019/06/Final_Report_Vol_1a-1.pdf.

173 **Eighty-four percent of American Indian:** André B. Rosay, "Violence Against American Indian and Alaska Native Women and Men," *National Institute of Justice Journal*, no. 277 (September 2016), www.ojp.gov/pdffiles1/nij/249822.pdf.

174 **1.2 times more:** Ibid.

174 **96 percent of sexual violence:** André B. Rosay, *Violence Against American Indian and Alaska Native Women and Men 2010 Findings from the National Intimate Partner and Sexual Violence Survey*, National Institute of Justice Research Report, May 2016, www.ojp.gov/pdffiles1/nij/249736.pdf.

174 **the stereotype of the "drunk Indian":** Robert J. Miller and Maril Hazlett, "The 'Drunken Indian': Myth Distilled into Reality Through Federal Indian Alcohol Policy," *Arizona State Law Journal* 28, no. 1 (1996), ssrn.com/abstract=1160478.

175 **"is an epistemology":** Lisa Grayshield, "Indigenous Ways of Knowing as a Philosophical Base for the Promotion of Peace and Justice in Counseling Education and Psychology," *Journal for Social Action in Counseling and Psychology* 2, no. 2 (Fall 2010), openjournals.bsu.edu/jsacp/article/view/311.

176 **Seven Grandfather Teachings:** "Seven Grandfather Teachings," Nottawaseppi Huron Band of the Potawatomi, nhbp-nsn.gov/seven-grandfather-teachings/.

176 **"In order for the whole to flourish":** Robin Wall Kimmerer, *Braiding Sweetgrass: Indigenous Wisdom, Scientific Knowledge and the Teachings of Plants* (Minneapolis: Milkweed Editions, 2013), 263.

177 **"the power to focus attention":** Ibid.

177 **sixty seconds of mindfulness:** Emma Chad-Friedman, Mojtaba Talaei-Khoei, David Ring, and Ana-Maria Vranceanu, "First Use of a Brief 60-Second Mindfulness Exercise in an Orthopedic Surgical Practice; Results from a Pilot Study," *Archives of Bone Joint Surgery* 5, no. 6 (2017): 400–5, www.ncbi.nlm.nih.gov/pmc/articles/PMC5736889/.

177 **reduce pain:** Bryany Cusens, Geoffrey B. Duggan, Kirsty Thorne, and Vidyamala Burch, "Evaluation of the Breathworks Mindfulness-Based Pain Management Programme: Effects on Well-Being and Multiple Measures of Mindfulness," *Clinical Psychology and Psychotherapy* 17, no. 1 (2009): 63–78.

177 **Mindfulness has been shown:** Shian-Ling Keng, Moria J. Smoski, and Clive J. Robins, "Effects of Mindfulness on Psychological Health: A Review of Empirical Studies," *Clinical Psychology Review* 31 (2011): 1041–56, doi.org/10.1016/j.cpr.2011.04.006.

177 **"arts can awaken people":** Thema Bryant, *Homecoming* (New York: Penguin Random House, 2022), 142.

177 **Traditional and Indigenous art:** Girija Kaimal and Asli Arslanbek, "Indigenous and Traditional Visual Artistic Practices: Implications for Art Therapy Clinical Practice and Research," *Frontiers in Psychology* 31 (2020), doi.org/10.3389/fpsyg.2020.01320.

177 **powerful tool of decolonization:** Lauren Elkin, "Recognition, at Last, After Decades Decolonizing Art," *New York Times,* October 15, 2021, www.nytimes.com/2021/10/15/arts/design/sutapa-biswas-art.html.

178 **Medicine Wheel teachings:** Hayley Zimak, "An Ontario Firekeeper Explains the Four Directions of the Medicine Wheel," Canadian Broadcasting Corporation, June 19, 2020, www.cbc.ca/news/canada/thunder-bay/four-directions-medicine-wheel-1.5615827.

179 **"civilization of the Mind":** Barlow, "Cyberspace."

179 **"global conversation of bits":** Ibid.

Chapter 11: Therapy

180 **Stan Twitter:** Tim Highfield, Stephen Harrington, and Axel Bruns, "Twitter as a Technology for Audiencing and Fandom," *Information, Communication and Society* 16, no. 3 (2013): 315–39, doi.org/10.1080/1369118X.2012.756053.

180 **bullying is the norm:** China Tony, "How Stan Twitter Is Turning into a

Community of Bullies," *Affinity*, December 29, 2016, affinitymagazine .us/2016/12/29/how-stan-twitter-is-turning-into-a-community-of-bullies.

181 **toppling Confederate statues:** Morgan Sung, "Here's How to *Hypothetically* Take Down a Racist Statue with Tips from an Archaeologist," *Mashable*, June 12, 2020, mashable.com/article/how-to-take-down-racist-statues-safely.

181 **began removing Confederate symbols:** Teo Armus, "'If We Don't Move It, They'll Take It Down': Protests Prompt U.S. Leaders to Remove Confederate Monuments," *Washington Post,* June 9, 2020, www.washingtonpost.com /nation/2020/06/09/confederate-monuments-protests-floyd.

181 **"smug little fascist":** Dinesh D'Souza (@DineshDSouza), Twitter, June 24, 2020, twitter.com/DineshDSouza/status/1275819522319147008.

183 **Harris navigated:** Alia E. Dastagir, "What Kamala Harris Put Up With," *USA Today,* October 8, 2020, www.usatoday.com/story/news/nation /2020/10/08/vp-debate-kamala-harris-and-sexism-racism-black-women -face/5924748002.

184 **White men founded:** Lillian Comas-Díaz and Edil Torres Rivera, eds., *Liberation Psychology: Theory, Method, Practice, and Social Justice* (Washington, DC: American Psychological Association, 2020), xv.

184 **steeped in some of the same values:** Ibid., 6.

184 **pathologizing women's pain:** Cecilia Tasca, Mariangela Rapetti, Mauro Giovanni Carta, and Bianca Fadda, "Women and Hysteria in the History of Mental Health," *Clinical Practice and Epidemiology in Mental Health* 8 (2012): 110, doi.org/10.2174/1745017901208010110.

184 **81 percent of active psychologists:** "Data Tool: Demographics of the U.S. Psychology Workforce," American Psychological Association, 2017, www.apa.org/workforce/data-tools/demographics.

185 **liberation psychology:** Comas-Díaz and Rivera, *Liberation Psychology,* 17–40.

185 **wrote the seminal text:** Paulo Freire, *Pedagogy of the Oppressed* (1970).

185 **"the master's tools":** Audre Lorde, "Age, Race, Class, and Sex: Women Redefining Difference," in *Sister Outsider: Essays and Speeches* (Berkeley, CA: Crossing Press, 1984), 116.

185 **"old blueprints of expectation and response":** Ibid., 115.

186 **Bonnie Burstow:** Julia Carmel, "Bonnie Burstow, Psychotherapist Who Rejected Psychiatry, Dies at 74," *New York Times,* January 31, 2020, www .nytimes.com/2020/01/31/health/bonnie-burstow-dead.html.

186 **"violence is absolutely integral":** Bonnie Burstow, *Radical Feminist Therapy: Working in the Context of Violence* (Thousand Oaks, CA: Sage Publications, 1992), xv.

186 **"a verbal journey of a witness":** Comas-Díaz and Rivera, *Liberation Psychology,* 133–48.

186 ***acompañamiento*:** Ibid., 93.

187 **cognitive behavioral therapy:** "What Is Cognitive Behavioral Therapy?," American Psychological Association, 2017, www.apa.org/ptsd-guideline /patients-and-families/cognitive-behavioral.

188 **break she took from writing:** Morgan Sung, "microdosing retirement," Rat. House, April 8, 2024, www.rat.house/p/microdosing-retirement.

188 **old post she made on Instagram:** Morgan Sung (@morgansung), Instagram, June 24, 2020.

Chapter 12: Fighting Back

191 **#SolidarityIsForWhiteWomen:** Susana Loza, "Hashtag Feminism, #SolidarityIsForWhiteWomen, and the Other #FemFuture," *Ada: A Journal of Gender, New Media, and Technology*, no. 5 (2014), awdflibrary.org/index.php ?p=show_detail&id=424.

191 **"disparaged by some":** Sarah J. Jackson, Moya Bailey, and Brooke Foucault Welles, *#HashtagActivism: Networks of Race and Gender Justice* (Cambridge, MA: MIT Press, 2020), 60.

191 **about her abortion:** Mikki Kendall, "Abortion Saved My Life," *Salon*, May 26, 2011, www.salon.com/2011/05/26/abortion_saved_my_life.

191 **saved her from a gunfight:** Mikki Kendall, *Hood Feminism: Notes from the Women That a Movement Forgot* (New York: Viking, 2020), 15.

191 **saved herself from an abusive marriage:** Ibid., 22–26.

195 **"Women are expected to":** Nina Jankowicz, *How to Be a Woman Online: Surviving Abuse and Harassment, and How to Fight Back* (New York: Bloomsbury Academic, 2022), 55.

195 **become more tolerant of it:** Robyn K. Mallett, Thomas E. Ford, and Julie A. Woodzicka, "Ignoring Sexism Increases Women's Tolerance of Sexual Harassment," *Self and Identity* 20, no. 7 (2021): 913–29, doi.org/10.1080/152 98868.2019.1678519.

195 **the goal to be respected:** Robyn K. Mallett and Kala J. Melchiori, "Goals Drive Responses to Perceived Discrimination," in *Confronting Prejudice and Discrimination: The Science of Changing Minds and Behaviors,* ed. R. K. Mallett and M. J. Monteith, 95–119 (Elsevier Academic Press, 2019), doi.org/10 .1016/B978-0-12-814715-3.00009-6.

197 **are treated as victims less:** maya finoh and jasmine Sankofa, "The Legal System Has Failed Black Girls, Women, and Non-Binary Survivors of Violence," ACLU, January 28, 2019, www.aclu.org/news/racial-justice/legal -system-has-failed-black-girls-women-and-non.

197 **"There is no set of years":** Christina Sharpe, *Ordinary Notes* (New York: Farrar, Straus & Giroux, 2023), 331.

197 **"the things Blacks have had":** Gwendolyn D. Pough, *Check It While I Wreck It: Black Womanhood, Hip-Hop Culture, and the Public Sphere* (Boston: Northeastern University Press, 2004), 16.

199 **wrote in a column for MSNBC:** André Brock, "What's Going to Happen to Black Twitter?" MSNBC, July 17, 2023, www.msnbc.com/opinion/msnbc -opinion/elon-musk-black-twitter-future-rcna93458.

200 **Jamilah Lemieux was targeted:** Demetria Irwin, "#StandWithJamilah: Black Twitter Rallies Behind Ebony Editor After Spat with RNC," *Grio*, thegrio .com/2014/03/31/standwithjamilah-black-twitter-rallies-behind-ebony -editor-after-spat-with-rnc.

200 **Black women online:** Mikki Kendall, *Hood Feminism: Notes from the Women That a Movement Forgot* (New York: Viking, 2020), 15.

200 **Anthea Butler:** Juan Thompson, "Online, Black Writers Confront Racist Backlash," *Intercept*, May 7, 2015, theintercept.com/2015/05/07/black-writers-confront-online-racism.

200 **Eve L. Ewing:** Keith Murphy, "Marvel Comics' Ironheart Writer Eve L. Ewing Is Ultimate Chicago Sports Homer," Andscape, January 4, 2009, andscape.com/features/marvel-comics-ironheart-writer-eve-l-ewing-is-ultimate-chicago-sports-homer.

200 **bystanders tend to evaluate confronters:** Alexander M. Czopp, "The Consequences of Confronting Prejudice," in *Confronting Prejudice and Discrimination: The Science of Changing Minds and Behaviors*, ed. Robyn K. Mallett and Margo J. Monteith, 201–21 (Elsevier Academic Press, 2019), doi.org/10.1016/B978-0-12-814715-3.00005-9.

200 **wrote in a 2022 blog post:** Emma Katz, "Domestic Abuse Survivors Who Fight Back," *Decoding Coercive Control with Dr. Emma Katz*, November 27, 2022, dremmakatz.substack.com/p/domestic-abuse-survivors-who-fight.

201 **"If the goal is to dismantle":** Whitney Phillips, *This Is Why We Can't Have Nice Things: Mapping the Relationship between Online Trolling and Mainstream Culture* (Cambridge, MA: MIT Press, 2015), 222.

202 **"afford to lose more voices":** Andrea Gibson, "The Cost of Call-Out Culture," *Things That Don't Suck,* June 30, 2022, andreagibson.substack.com/p/the-cost-of-call-out-culture.

202 **women who confront sexism:** Mallett, Ford, and Woodzicka, "Ignoring Sexism."

202 **women kept track of daily experiences:** Mindi Foster, "Everyday Confrontation of Discrimination: The Well-Being Costs and Benefits to Women over Time," *International Journal of Psychological Studies* 5, no. 3 (2013), doi.org/10.5539/ijps.v5n3p135.

203 **mental health guide:** Ana Maria Zellhuber Pérez and Juan Carlos Segarra Pérez, "A Mental Health Guide for Journalists Facing Online Violence," International Women's Media Foundation, November 2022, www.iwmf.org/wp-content/uploads/2022/12/Final_IWMF-Mental-health-guide.pdf.

204 **"feminist digilantism":** Emma A. Jane, "Online Misogyny and Feminist Digilantism," *Continuum* 30, no. 3 (2016): 284–97, doi.org/10.1080/10304312.2016.1166560.

Chapter 13: A Humanity That Isn't Up for Debate

206 **punctures a waterbed:** "Edward Scissorhands 'Water Bed' Clip," August 20, 2011, www.youtube.com/watch?v=DLVJjvyguHM.

206 **"an uncommonly gentle man":** *Edward Scissorhands*, Taglines, IMDb, www.imdb.com/title/tt0099487/taglines.

207 **She filed for divorce:** Yanan Wang, "From Dog Fight to Divorce: Johnny Depp and Amber Heard," *Washington Post,* May 26, 2016, www.washingtonpost.com/news/morning-mix/wp/2016/05/26/from-dog-fight-to-divorce-johnny-depp-and-amber-heard.

207 **restraining order:** Emily Yahr, "Amber Heard Files Restraining Order Against Johnny Depp After Alleged Domestic Violence," *Washington Post,* May 27, 2016, www.washingtonpost.com/news/arts-and-entertainment/wp/2016/05

/27/amber-heard-files-restraining-order-against-johnny-depp-after-alleged -domestic-violence.

207 **pictures of her from the courthouse:** Maane Khatchatourian, "Amber Heard Granted Domestic Violence Restraining Order Against Johnny Depp," *Variety*, May 27, 2016, variety.com/2016/film/news/amber-heard-johnny -depp-domestic-violence-restraining-oder-1201784462.

207 **small bruise:** Jodi Guglielmi, "Amber Heard Seen with Black Eye Leaving L.A. Courthouse After Domestic Violence Claim Against Johnny Depp," *People*, May 27, 2016, people.com/movies/amber-heard-seen-with-black-eye -leaving-l-a-courthouse.

207 **security guard:** Shauna Snow and Aleene Macminn, "Legal File," *Los Angeles Times,* September 11, 1989, www.latimes.com/archives/la-xpm-1989-09-11 -ca-1377-story.html.

207 **damages to a hotel room:** Nadine Brozan, "Chronicle," *New York Times,* September 14, 1994, www.nytimes.com/1994/09/14/style/chronicle-236659 .html.

207 **struggles with alcohol and drugs:** Emily Yahr and Sonia Rao, "Cross-examination Spotlights Depp's Profane Texts, Drug Use," *Washington Post,* April 21, 2022, www.washingtonpost.com/arts-entertainment/2022/04/21 /depp-heard-trial-cross-examine.

207 **"substantially true":** Elahe Izadi and Sarah Ellison, "Why Johnny Depp Lost His Libel Case in the U.K. But Won in the U.S.," *Washington Post,* June 1, 2022, www.washingtonpost.com/media/2022/06/01/johnny-depp-libel-law-uk-us.

208 **horrific details of abuse:** Alia E. Dastagir, "Amber Heard Says She's a Victim, But the Public Made Her a Villain. Experts Say It's a Dangerous Moment for Domestic Violence," *USA Today,* May 10, 2022, www.usatoday.com/story/life /health-wellness/2022/05/10/amber-heard-tiktoks-johnny-depp-trial -triggering-abuse-victims/9707944002.

208 **sexually assaulted her:** Naledi Ushe, "Amber Heard Details Graphic Account of Alleged Sexual Assault by Johnny Depp: 'I Was Scared,'" *USA Today,* May 5, 2022, www.usatoday.com/story/entertainment/celebrities/2022/05/05 /amber-heard-testimony-day-2-johnny-depp-trial-libel-case/9656638002.

208 **beat her:** Travis M. Andrews and Sonia Rao, "Heard Continues Testimony, Describing 'Pattern' of Depp's Abuse," *Washington Post,* May 5, 2022, www .washingtonpost.com/arts-entertainment/2022/05/05/johnny-depp-amber -heard-trial-abuse.

208 **berated her:** Pilar Melendez, "'Shut Up, Fat Ass': Audio Reveals Aftermath of Johnny Depp Allegedly Putting Cigarette Out on Amber Heard," *Daily Beast,* April 25, 2022, www.thedailybeast.com/johnny-depp-calls-amber-heard-a -fat-ass-after-alleged-cigarette-attack-in-audio-played-in-defamation-trial.

208 **controlled her:** Andrews and Rao, "Heard Continues Testimony."

208 **put cigarettes out:** Melendez, "'Shut Up, Fat Ass.'"

208 **bisexual woman:** Alia E. Dastagir, "What the Amber Heard, Johnny Depp Trial Didn't Cover: The Violence Bisexual Women Face," *USA Today,* June 15, 2022, www.usatoday.com/story/life/health-wellness/2022/06/15/amber -heard-johnny-depp-violence-bisexual-women/7621083001.

208 **denied the opportunity to present evidence:** Claire Lampen, "Amber Heard Releases Therapy Notes Excluded from Trial," *Cut*, June 19, 2022, www.thecut.com/2022/06/amber-heard-doesnt-blame-the-jury-for-johnny-depp-verdict.html.

208 **therapy notes:** Kylie Cheung, "Celebrities Are Unliking Johnny Depp's Post-Trial Instagram Post After New Court Documents Were Unsealed," *Jezebel*, August 5, 2022, www.jezebel.com/celebrities-are-unliking-johnny-depp-s-post-trial-insta-1849377565.

208 **text messages from a Depp employee:** Marlow Stern, "Unsealed Johnny Depp v. Amber Heard Court Documents Reveal Shocking New Claims," *Daily Beast,* July 31, 2022, www.thedailybeast.com/unsealed-docs-from-johnny-depp-v-amber-heard-defamation-trial-contain-shocking-new-claims.

208 **"mutual abuse":** Ben Finley, "Therapist: Depp and Heard Had Relationship of 'Mutual Abuse,'" Associated Press, April 14, 2022, apnews.com/article/entertainment-amber-heard-johnny-depp-lawsuits-arts-and-fe315bb74c8d89303545e366aabff89f.

208 **not used in the field:** Alia E. Dastagir, "Amber Heard Has Yet to Take the Stand. But on Social Media, Johnny Depp Has Already Won," *USA Today,* April 29, 2022, www.usatoday.com/story/life/health-wellness/2022/04/29/johnny-depp-amber-heard-trial-tiktok-videos-domestic-abuse/9583038002.

208 **dynamics of intimate partner violence:** Alia E. Dastagir, "Juror in Amber Heard Case Said She Wasn't 'Believable.' What Experts in Domestic and Sexual Violence Say About Believability," *USA Today,* June 16, 2022, www.usatoday.com/story/life/health-wellness/2022/06/16/amber-heard-juror-says-she-wasnt-believable-trauma-experts-weigh-in/7651713001.

208 **they didn't believe her:** Alia E. Dastagir, "Amber Heard, Johnny Depp and Who We Choose to Believe," *USA Today,* May 23, 2022, www.usatoday.com/story/life/health-wellness/2022/05/23/amber-heard-johnny-depp-trial-biases-tiktok-making-judgements/9887926002.

209 **who were not sequestered:** Alia E. Dastagir, "Why We Didn't See Amber Heard Coming and What It Might Mean for Other Women Who Allege Abuse," *USA Today,* June 2, 2022, www.usatoday.com/story/life/health-wellness/2022/06/02/after-amber-heard-we-asking-wrong-questions-metoo/7487231001.

209 **O.J. Simpson trial:** David Margolick, "Not Guilty: The Overview; Jury Clears Simpson in Double Murder; Spellbound Nation Divides on Verdict," *New York Times,* October 4, 1995, www.nytimes.com/1995/10/04/us/not-guilty-overview-jury-clears-simpson-double-murder-spellbound-nation-divides.html.

209 **"You won't ever know":** Johanna Fateman and Amy Scholder, ed., *Last Days at Hot Slit: The Radical Feminism of Andrea Dworkin* (Cambridge, MA: MIT Press, 2019), 289.

209 **Rebecca remembered Monica Lewinsky:** "Monica Lewinsky: Emerging from 'the House of Gaslight' in the Age of #MeToo," *Vanity Fair*, February 25, 2018, www.vanityfair.com/news/2018/02/monica-lewinsky-in-the-age-of-metoo.

209 **a small group:** "Mega List of Pro-Amber Heard Twitter Accounts to Follow,"

Reddit, www.reddit.com/r/DeppDelusion/comments/wq29vb/mega_list_of_proamber_heard_twitter_accounts_to.

209 **significantly coordinated:** "Targeted Trolling and Trend Manipulation: How Organized Attacks on Amber Heard and Other Women Thrive on Twitter," *Bot Sentinel*, July 18, 2022, botsentinel.com/reports/documents/amber-heard/report-07-18-2022.pdf.

209 **"episodically":** Kat Tenbarge, "A Pro-Johnny Depp YouTuber Was Sent a Cease-and-Desist After She Targeted an Employee of Amber Heard's PR Team," NBC News, August 12, 2022, www.nbcnews.com/tech/internet/-johnny-depp-youtuber-was-sent-cease-desist-targeted-employee-amber-he-rcna42741.

209 **specifically admitted to communicating:** "Johnny Depp's Attorney Speaks to Domestic Abuse Claims," Court TV, May 19, 2022, www.facebook.com/courttv/videos/johnny-depps-attorney-speaks-to-domestic-abuse-claims-court-tv/1225693334901755/?_rdr.

209 **grow their platforms:** Tenbarge, "Depp YouTuber."

210 **"You deserve a community":** Thema Bryant, *Homecoming* (New York: Penguin Random House, 2022), 217.

210 **social support:** Shelley E. Taylor, "Social Support: A Review," in *The Oxford Handbook of Health Psychology*, ed. Howard S. Friedman (New York: Oxford University Press, 2011), 189–214.

211 **less likely to develop PTSD:** Daniel F. Gros et al., "Relations Among Social Support, PTSD Symptoms, and Substance Use in Veterans," *Psychology of Addictive Behaviors* 30, no. 7 (2016): 764, doi.org/10.1037/adb0000205.

211 **peer support:** Susan E. McGregor, Viktorya Vilk, and Jeje Mohamed, "The Power of Peer Support," PEN America, April 2024, pen.org/report/peer-support.

211 **evidence-based practice:** Shery Mead, David Hilton, and Laurie Curtis, "Peer Support: A Theoretical Perspective," *Psychiatric Rehabilitation Journal* 25, no. 2 (2001): 134–41, doi.org/10.1037/h0095032.

211 **as a response to online violence:** "Welcome to Right to Be's Storytelling Platform," *Right to Be*, stories.righttobe.org.

212 **"People cannot feel safe alone":** Judith L. Herman, *Truth and Repair: How Trauma Survivors Envision Justice* (New York: Basic Books, 2023), 8.

212 **clarify our experiences:** Stephanie A. Shields, "The Politics of Emotion in Everyday Life: 'Appropriate' Emotion and Claims on Identity," *Review of General Psychology* 9, no. 1 (2005): 3–15, doi.org/10.1037/1089-2680.9.1.3.

213 **aid us in the process of reporting:** Viktorya Vilk and Kat Lo, "Shouting into the Void: Why Reporting Abuse to Social Media Platforms Is So Hard and How to Fix It," PEN America, June 29, 2023, pen.org/report/shouting-into-the-void.

213 **when a person sees a stranger:** M. L. Meyer et al., "Empathy for the Social Suffering of Friends and Strangers Recruits Distinct Patterns of Brain Activation," *Social Cognitive and Affective Neuroscience* 8, no. 4 (2013): 446–54, doi.org/10.1093/scan/nss019.

213 **involved in mentalizing:** C. L. Masten, S. A. Morelli, and N. I. Eisenberger,

"An fMRI Investigation of Empathy for 'Social Pain' and Subsequent Prosocial Behavior," *Neuroimage* 55, no. 1 (2011): 381–88, doi.org/10.1016/j.neuroimage.2010.11.060.

214 **"One thing I love":** Bridget Todd, "Going Boldly with Blair Imani," *There Are No Girls on the Internet* (podcast), Episode 009, August 18, 2020, www.tangoti.com/episode-9.

214 **Adam Conover retweeted her parody:** Pamela Ross, "Spotlight: Kelsey Caine's Irreverent Satire," WICF Daily, n.d., daily.wicf.com/spotlight-kelsey-caines-irreverent-satire.

214 **"When we have to fight for":** Sara Ahmed, *Living a Feminist Life* (Durham, NC: Duke University Press, 2017), 176.

215 **men's rights activists went mainstream:** Eddie Kim, "Johnny Depp, Men's Rights Hero," *MEL Magazine*, n.d., melmagazine.com/en-us/story/amber-heard-johnny-depp-trial.

216 **"one of the worst cases":** "Targeted Trolling and Trend Manipulation: How Organized Attacks on Amber Heard and Other Women Thrive on Twitter," *Bot Sentinel*, July 18, 2022, botsentinel.com/reports/documents/amber-heard/report-07-18-2022.pdf.

216 **"single-purpose hate accounts":** "Coordinated Hate Campaign Targeting Harry and Meghan, Duke and Duchess of Sussex," *Bot Sentinel*, January 18, 2022, botsentinel.com/reports/documents/duke-and-duchess-of-sussex/report-01-18-2022.pdf.

216 **more than 10 million views:** Kat Tenbarge, "YouTube Creators Are Pivoting Their Videos to Depp v. Heard Content and Raking in Millions of Views," NBC News, May 13, 2022, www.nbcnews.com/tech/internet/amber-heard-trial-depp-youtube-content-platform-algorithm-tiktok-rcna28016.

216 **Depp sued Heard for $50 million:** Travis M. Andrews and Emily Yahr, "Jury Rules Actors Johnny Depp and Amber Heard Defamed Each Other," *Washington Post,* June 1, 2022, www.washingtonpost.com/arts-entertainment/2022/06/01/depp-wins-jury-verdict.

217 **"a public figure representing domestic abuse":** "Amber Heard: I Spoke Up Against Sexual Violence—And Faced Our Culture's Wrath. That Has to Change," *Washington Post,* December 18, 2018, www.washingtonpost.com/opinions/ive-seen-how-institutions-protect-men-accused-of-abuse-heres-what-we-can-do/2018/12/18/71fd876a-02ed-11e9-b5df-5d3874f1ac36_story.html.

217 **Depp had promised would happen:** Dastagir, "Why We Didn't See Amber Heard Coming."

217 **"I am harassed, humiliated, threatened":** Julia Jacobs, "Amber Heard Describes Impact of Online Attacks: 'I'm a Human Being,'" *New York Times,* May 26, 2022, www.nytimes.com/2022/05/26/arts/amber-heard-johnny-depp-harassment.html.

218 **"The victim":** Judith L. Herman, *Trauma and Recovery: The Aftermath of Violence—From Domestic Abuse to Political Terror* (New York: Basic Books, 1992), 15.

219 **the seven-thousand-word narrative:** Alia E. Dastagir, "'The Most Intense

Violation of My Life': A Beloved Camp, a Lost Boy and the Lifelong Impact of Child Sexual Trauma," *USA Today,* January 10, 2022, www.usatoday.com/in-depth/life/health-wellness/2022/01/10/summer-camp-child-sexual-abuse/8650964002.

219 **science of pedophilia:** Alia E. Dastagir, "The Complicated Research Behind Pedophilia," *USA Today,* January 10, 2022, www.usatoday.com/story/life/health-wellness/2022/01/10/pedophiles-pedophilia-sexual-disorder/8768423002.

219 **"a sprawling spiderweb":** "What You Need to Know About QAnon," Southern Poverty Law Center, October 27, 2020, www.splcenter.org/hatewatch/2020/10/27/what-you-need-know-about-qanon.

219 **has indicated he believes them:** David Klepper and Ali Swenson, "Trump Openly Embraces, Amplifies QAnon Conspiracy Theories," Associated Press, September 16, 2022, apnews.com/article/technology-donald-trump-conspiracy-theories-government-and-politics-db50c6f709b1706886a876ae6ac298e2.

219 **tweeted a screenshot of it:** Donald Trump, Jr. (@DonaldJTrumpJr), Twitter, January 11, 2022, twitter.com/DonaldJTrumpJr/status/1480976148725223424.

220 **Fox News:** Lindsay Kornick, "USA Today Torched for Promoting 'Complicated' Study on Pedophilia," Fox News, January 11, 2022, www.foxnews.com/media/usa-today-blasted-promoting-complicated-pedophilia-study.

220 **Breitbart:** Paul Bois, "USA Today Tries to Destigmatize Pedophilia, Fails Miserably," *Breitbart,* January 11, 2022, www.breitbart.com/politics/2022/01/11/usa-today-tries-to-destigmatize-pedophilia-fails-miserably.

220 **the *Daily Mail*:** Gina Martinez, "USA Today Faces Backlash and Deletes Series of Tweets Which 'Normalize' PEDOPHILIA by Claiming It Is 'Misunderstood' and a Condition 'Determined in the Womb,'" *Daily Mail,* January 12, 2022, www.dailymail.co.uk/news/article-10395407/USA-Today-faces-backlash-deletes-series-tweets-normalizes-pedophilia.html.

220 **online harassment of women scholars:** J. Hodson, C. Gosse, G. Veletsianos, and S. Houlden, "I Get By with a Little Help from My Friends: The Ecological Model and Support for Women Scholars Experiencing Online Harassment," *First Monday* 23, no. 8 (2018), doi.org/10.5210/fm.v23i8.9136.

221 **statement of support:** "Best Practices for Employers," *Online Harassment Field Manual,* PEN America, onlineharassmentfieldmanual.pen.org/best-practices-for-employers.

Conclusion

225 **Republicans and Democrats held similar views:** Emily A. Vogels, "Partisans in the U.S. Increasingly Divided on Whether Offensive Content Online Is Taken Seriously Enough," Pew Research Center, October 8, 2020, pewresearch.org/short-reads/2020/10/08/partisans-in-the-u-s-increasingly-divided-on-whether-offensive-content-online-is-taken-seriously-enough.

226 **"We can't separate lives":** María C. Lugones and Elizabeth V. Spelman, "Have

We Got a Theory for You! Feminist Theory, Cultural Imperialism and the Demand for 'The Woman's Voice,'" *Women's Studies International Forum* 6, no. 6 (1983): 573–81, doi.org/10.1016/0277-5395(83)90019-5.

226 **"with time, resources, and will":** Viktorya Vilk and Kat Lo, "Shouting into the Void: Why Reporting Abuse to Social Media Platforms Is So Hard and How to Fix It," PEN America, June 29, 2023, pen.org/report/shouting-into-the-void.

227 **"exclusion built into its design":** Ellen Pao, *Reset: My Fight for Inclusion and Lasting Change* (New York: Spiegel & Grau, 2017), 240.

227 **made it a priority:** Emily Chang, *Brotopia: Breaking Up the Boys' Club of Silicon Valley* (New York: Portfolio/Penguin, 2018), 244.

227 **attacked by trolls:** Ellen Pao, "Former Reddit CEO Ellen Pao: The Trolls Are Winning the Battle for the Internet," *Washington Post,* July 16, 2015, washingtonpost.com/opinions/we-cannot-let-the-internet-trolls-win/2015/07/16/91b1a2d2-2b17-11e5-bd33-395c05608059_story.html.

227 **80 percent drop in hate speech:** Eshwar Chandrasekharan et al., "You Can't Stay Here: The Efficacy of Reddit's 2015 Ban Examined Through Hate Speech," *Proceedings of the ACM on Human-Computer Interaction* 1, no. CSCW (2017), doi.org/10.1145/3134666.

227 **changes to Section 230:** "Section 230," Electronic Frontier Foundation, www.eff.org/issues/cda230.

227 **disproportionately impact marginalized communities:** Letter from 70+ Human Rights and Social Justice Groups Opposing Repeal of or Overbroad Changes to Section 230, *Fight for the Future,* January 27, 2021, www.fightforthefuture.org/news/2021-01-27-letter-from-70-human-rights-and-social-justice.

227 **silence social and racial justice movements:** Evan Greer and Lia Holland, "Section 230 Is a Last Line of Defense for Abortion Speech Online," *Wired,* June 29, 2022, www.wired.com/story/section-230-is-a-last-line-of-defense-for-abortion-speech-online.

228 **their paper on the practice:** Bianca Cepollaro, Maxime Lepoutre, and Robert Mark Simpson, "Counterspeech," *Philosophy Compass* 18, no. 1 (2023): e12890, doi.org/10.1111/phc3.12890.

228 **In previous eras of communication:** Sofia Rüdiger and Daria Dayter, eds., *Corpus Approaches to Social Media* (Amsterdam: John Benjamins Publishing, 2020).

229 **trauma-informed technology:** Trauma-Informed Technology, www.traumainformedtech.com/learning.

229 **relive past trauma:** Carol F. Scott and Melissa Eggleston, "Inclusive by Design: Use a Trauma-Informed Approach," *User Experience,* December 19, 2023, uxpamagazine.org/inclusive-by-design-use-a-trauma-informed-approach.

229 **"you only code as well as":** Sydette Harry, "You Only Code as Well as You Listen," *Code for America*, August 16, 2019, codeforamerica.org/news/you-only-code-as-well-as-you-listen.

231 **"violence is not a subject":** Jacqueline Rose, *On Violence and On Violence Against Women* (New York: Farrar, Straus & Giroux, 2021), 359.

ABOUT THE AUTHOR

ALIA DASTAGIR is an award-winning former reporter for *USA Today* who frequently covers gender and mental health. Dastagir was one of eight U.S. recipients of the prestigious Rosalynn Carter Fellowships for Mental Health Journalism. She won a first-place National Headliner Award for a series on suicide and was the first winner of the American Association of Suicidology's Public Service Journalism Award. The Media Awards Committee for the Council on Contemporary Families named her story on America's lack of affordable childcare the winner for Outstanding Media Coverage of Family Issues. Dastagir is currently pursuing an MFA in creative writing at NYU, where she is an Axinn fellow.